SpringerBriefs in Mathematics

SpringerBriefs present concise summaries of cutting-edge research and practical applications across a wide spectrum of fields. Featuring compact volumes of 50 to 125 pages, the series covers a range of content from professional to academic. Briefs are characterized by fast, global electronic dissemination, standard publishing contracts, standardized manuscript preparation and formatting guidelines, and expedited production schedules.

Typical topics might include:

- A timely report of state-of-the art techniques
- A bridge between new research results, as published in journal articles, and a contextual literature review
- A snapshot of a hot or emerging topic
- An in-depth case study
- A presentation of core concepts that students must understand in order to make independent contributions

SpringerBriefs in Mathematics showcases expositions in all areas of mathematics and applied mathematics. Manuscripts presenting new results or a single new result in a classical field, new field, or an emerging topic, applications, or bridges between new results and already published works, are encouraged. The series is intended for mathematicians and applied mathematicians. All works are peer-reviewed to meet the highest standards of scientific literature.

Titles from this series are indexed by Scopus, Web of Science, Mathematical Reviews, and zbMATH.

Mark R. Budden

Star-Critical Ramsey Numbers for Graphs

Springer

Mark R. Budden
Department of Mathematics and Computer Science
Western Carolina University
Cullowhee, NC, USA

ISSN 2191-8198 ISSN 2191-8201 (electronic)
SpringerBriefs in Mathematics
ISBN 978-3-031-29980-3 ISBN 978-3-031-29981-0 (eBook)
https://doi.org/10.1007/978-3-031-29981-0

Mathematics Subject Classification: 05C55, 05C15, 05C35, 05D10

This Springer imprint is published by the registered company Springer Nature Switzerland AG
The registered company address is: Gewerbestrasse 11, 6330 Cham, Switzerland

This book is dedicated to my wife, Beth, for her constant encouragement, support, and tolerance for the time I commit to mathematics.

Preface

Star-critical Ramsey numbers were originally introduced by Jonelle Hook in her dissertation [47] in 2010, working under the direction of her advisor, Garth Isaak. In their short existence, star-critical Ramsey numbers have earned a position of prominence in the Ramsey theory of graphs. For a collection of graphs, the star-critical Ramsey number serves as a measure of the strength of the corresponding Ramsey number, hinting at the connectivity of the underlying graphs. Determining the exact value of a star-critical number involves a somewhat similar process to the determination of the corresponding Ramsey number, allowing standard techniques in Ramsey theory to be applied to this related topic.

In this monograph, I have tried to capture the current state-of-knowledge on star-critical Ramsey numbers and their generalizations. Emphasis has been placed on developing standard proof techniques and constructions, while offering numerous directions for future research. Tables of known results are also provided for quick reference. It is hoped that both students and experienced researchers will find this book to be a valuable resource.

I owe a great deal of gratitude to my former research students, as these collaborations served as the primary motivation for the creation of this book. In particular, my work with Elijah DeJonge ([4] and [5]) and Zoulaiha Daouda [3] on star-critical Ramsey numbers appears in this book. Appreciation is also due to Garth Isaak for informing me of the connection between star-critical Ramsey numbers and deleted edge numbers and Jonelle Hook for sharing a preliminary version of her survey article [49] with me. Their foundational work on star-critical Ramey numbers continues to motivate many Ramsey theorists.

Cullowhee, NC, USA   Mark R. Budden
January 2023

Contents

List of Figures

Chapter 1
Introduction to Multicolor Star-Critical Ramsey Numbers

1.1 Graph-Theoretic Background

At its core, Ramsey theory is the study of set partitions. One wishes to determine how large a set must be to guarantee that a partition of a given size possesses desireable properties. In the Ramsey theory of graphs, the primary focus is on edge colorings of complete graphs, identifying how large their orders must be to guarantee the existence of certain monochromatic subgraphs. Star-critical Ramsey theory serves to refine the notion of Ramsey numbers. Describing this theory requires a detailed account of the history and background surrounding Ramsey numbers.

We begin by reviewing the standard definitions and notations from graph theory used throughout this text. Background from Ramsey theory is also included, with references provided for those wishing to learn more about this ubiquitous area of mathematics. For basic references in graph theory, the reader is encouraged to consult Chartrand et al.'s book [15] (which includes a section on Ramsey theory), the classical text [44] by Harary, and the book by West [89]. Introductions to various aspects of Ramsey theory can be found in [35, 36, 56, 57, 59, 60, 71, 72, 79], and [82], and researchers seeking open problems in Ramsey theory may consult [91].

A *graph* $G = (V(G), E(G))$ consists of a nonempty vertex set $V(G)$ and a set of edges $E(G)$, consisting of 2-element subsets of $V(G)$. We write xy in place of the edge $\{x, y\} \in E(G)$. It is assumed that all graphs considered here are simple (i.e., loops and multiedges are not allowed). The number of vertices contained in a graph G, denoted $|V(G)|$, is called the *order of* G and the number of edges in G, denoted $|E(G)|$, is called the *size* of G. The *complement* of G, denoted $\overline{G}$, is the graph with vertex set $V(\overline{G}) = V(G)$ and edge set

$$E(\overline{G}) = \{xy \mid xy \notin E(G)\}.$$

© The Author(s), under exclusive license to Springer Nature Switzerland AG 2023
M. R. Budden, *Star-Critical Ramsey Numbers for Graphs*, SpringerBriefs in
Mathematics, https://doi.org/10.1007/978-3-031-29981-0_1

Two graphs G and G' are called *isomorphic* if there exists a bijection $f : V(G) \longrightarrow V(G')$ (called an *isomorphism*) such that vertices x and y are adjacent in G if and only if $f(x)$ and $f(y)$ are adjacent in G'.

Within a graph G, vertices that are contained in a common edge, and edges that have vertices in common, are called *adjacent*. If a vertex is contained in an edge, we say the vertex and edge are *incident*. The *degree* of a vertex $x \in V(G)$, denoted $\deg_G(x)$, is the number of edges incident with x. The *minimum degree* and *maximum degree* of a graph G are denoted by $\delta(G)$ and $\Delta(G)$, respectively:

$$\delta(G) := \min\{\deg_G(x) \mid x \in V(G)\} \quad \text{and} \quad \Delta(G) := \max\{\deg_G(x) \mid x \in V(G)\}.$$

A graph H is called a *subgraph* of G if $V(H) \subseteq V(G)$ and $E(H) \subseteq E(G)$. If $S \subseteq V(G)$, then the *subgraph of G induced by S*, denoted $G[S]$, has vertex set S and edge set

$$E(G[S]) = \{xy \in E(G) \mid x, y \in S\}.$$

The *complete graph of order p* vertices, denoted K_p, is the graph with p vertices in which every distinct pair of vertices are adjacent. We will often consider a complete graph of order p with a single edge removed, which is denoted $K_p - e$. A *path of order n*, denoted P_n, is a sequence of distinct vertices $x_1 x_2 \cdots x_n$ such that every consecutive pair of vertices form an edge (i.e., $x_i x_{i+1}$ is an edge for all $1 \le i \le n-1$). Here, we say that P_n has *length $n-1$* (the number of edges in P_n). Given a path $x_1 x_2 \cdots x_n$, if $x_n x_1$ is also an edge, then $x_1 x_2 \cdots x_n x_1$ is called a *cycle of length n*, and is denoted by C_n.

A graph is called *connected* if every distinct pair of vertices are contained within some subgraph that is isomorphic to a path. If a graph is not connected, its largest connected subgraphs are called its *connected components*. The *connectivity* of a graph G, denoted $\kappa(G)$, is the minimum number of points whose deletion results in a disconnected graph or a single vertex. If a graph is disconnected, then its connectivity is 0. If a graph G satisfies $\kappa(G) = 1$, then a vertex whose removal disconnects G is called a *cut vertex*. In the case of the complete graph K_n, deleting vertices never results in a disconnected graph, but removing $n-1$ of the vertices results in a single vertex. It follows that $\kappa(K_n) = n-1$. In the cases of paths and cycles, it is easily confirmed that $\kappa(P_n) = 1$ for all $n \ge 2$ and $\kappa(C_n) = 2$ for all $n \ge 3$. A graph G is called *k-connected* if $\kappa(G) \ge k$.

A subset of the vertex set of a graph G is called an *independent set of vertices* if no two of its vertices are adjacent in G. The maximum cardinality of an independent set of vertices in G is called the *independence number* of G, and is denoted $\alpha(G)$. Note that if $\alpha(G) = n$, then $\overline{G}$ contains a subgraph isomorphic to K_n, but no subgraph isomorphic to K_{n+1}.

A standard method for studying partitions of the vertex set or edge set of a graph is through the coloring of elements in such a set. In this sense, the colors correspond with the subsets of a partition. A *proper vertex coloring* of a graph G is an assignment of colors to the elements in $V(G)$ such that no two adjacent vertices receive the same color. Each color class of a proper vertex coloring of a graph is an independent sets of vertices. The minimum number of colors required to properly vertex color a graph G is called the *chromatic number* of G, and is denoted $\chi(G)$. Among all proper vertex colorings of a graph G using exactly $\chi(G)$ colors, the cardinality of the color class containing the fewest number of colors is called the *chromatic surplus* of G and is denoted $s(G)$. A simple consequence of this definition is that no proper vertex coloring of G using exactly $\chi(G)$ colors has a color class that uses fewer than $s(G)$ colors.

The colorings that will serve us most in the study of Ramsey theory are edge colorings. A *t-coloring* of a graph G is a function $c : E(G) \longrightarrow \{1, 2, \ldots, t\}$. Here, we think of the elements in $\{1, 2, \ldots, t\}$ as being "colors," and for small values of t, we often opt to use red, blue, green, etc. in place of the numbers $1, 2, \ldots, t$. In general, it is not assumed that c is surjective, so a graph may be referred to as being t-colored, even if its edges are assigned fewer than t colors. Given a t-coloring c of a graph G, we say that G contains a copy of H in color i if there exists some subgraph of G that is isomorphic to H in which all of its edges are assigned color i under c.

The most common value of t that we will encounter is $t = 2$, in which case, we may refer to a red/blue coloring of G in order to emphasize the specific colors being used. In this case, the subgraph of G spanned by the red edges will be denoted by G_R. If x is any vertex in G, we denote by $N_R(x)$ and $\deg_R(x)$ the *red neighborhood* of x and the *red degree* of x in G_R, respectively, We define G_B, $N_B(x)$, and $\deg_B(x)$ using the color blue, analogously.

If $G_1, G_2, \ldots G_t$ are graphs, then the *Ramsey number* $r(G_1, G_2, \ldots, G_t)$ is defined to be the least natural number p such that every t-coloring of K_p contains a copy of G_i in color i for some $1 \leq i \leq t$. From this definition, it is easily confirmed that

$$r(K_1, G_2, G_3, \ldots, G_t) = 1 \quad \text{and} \quad r(K_2, G_2, G_3, \ldots, G_t) = r(G_2, G_3, \ldots, G_t),$$

for all graphs $G_2, G_3, \ldots, G_t$. If $p = r(G_1, G_2, \ldots, G_t)$, then any t-coloring of K_{p-1} that avoids a monochromatic copy of G_i in color i, for all $1 \leq i \leq t$, is called a *critical coloring* for $(G_1, G_2, \ldots, G_t)$. We denote the set of all critical colorings for $(G_1, G_2, \ldots, G_t)$ by $\mathrm{Crit}(G_1, G_2, \ldots, G_t)$. We may often refer to an element of $\mathrm{Crit}(G_1, G_2, \ldots, G_t)$ as a "graph," but it should be understood that it a t-coloring of K_{p-1}.

In many of the graph constructions described in this text, the concept of replacing a vertex with a copy of another graph will be used. If G and G' are two graphs and v is a vertex in G, then the graph formed by replacing v with G' has vertex set $(V(G) - \{v\}) \cup V(G')$. Its edge set consists of the edges in $G[V(G) - \{v\}]$, $E(G')$, and

$$\{xy \mid x \in V(G) - \{v\}, \ y \in V(G'), \text{ and } xv \in E(G)\}.$$

In this case, G is called the *base graph* and G' is a *block*.

Many of the constructions we will consider involve operations on graphs. Let G_1 and G_2 be graphs. Then the *(disjoint) union* of G_1 and G_2, denoted $G_1 \cup G_2$, is the graph with vertex set $V(G_1) \cup V(G_2)$ and edge set $E(G_1) \cup E(G_2)$. The union of n copies of G will be denoted nG. The *join* of G_1 and G_2, denoted $G_1 + G_2$ has vertex set $V(G_1) \cup V(G_2)$ and edge set

$$E(G_1) \cup E(G_2) \cup \{uv \mid u \in V(G_1) \text{ and } v \in V(G_2)\}.$$

In other words, $G_1 + G_2$ is formed by replacing one vertex in a K_2 with G_1 and the other vertex with G_2. Note that some authors (in fact, many of those cited in this text) use $+$ for the union and $\vee$ for the join of graphs.

For $n \geq 2$, an *n-partite* graph is a graph in which its vertex set can be partitioned into nonempty subsets $V_1, V_1, \ldots, V_n$ (called *partite sets*) such that every edge contains vertices from two distinct partite sets. The *complete n-partite graph* $K_{k_1,k_2,\ldots,k_n}$ is the complement of $K_{k_1} \cup K_{k_2} \cup \cdots \cup K_{k_n}$. A 2-partite graph is referred to as *bipartite* and the bipartite graph $K_{1,n}$ is called a *star*. When $k_1 = k_2 = \cdots = k_n$, we write $K_n(k_1)$ for the corresponding complete n-partite graph.

Given a graph G, a subgraph G' *spans* G if $V(G') = V(G)$. Any spanning subgraph of a graph G is called a *factor* of G and a k-regular factor (one in which all of its vertices have degree k) is called a *k-factor*. A 1-factor is also called a *perfect matching* and a 2-factor is a spanning cycle. The graph G is said to be the *sum* of factors $G_1, G_2, \ldots G_t$ if G is the line-disjoint union of $G_1, G_2, \ldots, G_t$. If G is the sum of k-factors, then their union is called a *k-factorization* of G and we say that G is *k-factorable*.

A 2-factor (or spanning cycle) of a graph G is also called a *Hamiltonian cycle*. If G has a 2-factor, we say that G is *Hamiltonian*. A *Hamiltonian path* in a graph G is a path that includes all vertices in G. If a graph has a Hamiltonian path, then it is necessarily connected. In 1963, Ore [74] defined a graph G to be *Hamiltonian-connected* if there exists a Hamiltonian path between every distinct pair of vertices in $V(G)$.

A connected graph that does not contain any cycles is called a *tree*. A vertex of degree 1 in a tree is called a *leaf*. It is well-known that a graph is a tree if and only if it is *minimally connected* (i.e., it is connected, but the deletion of any edge, while retaining all vertices, results in a disconnected graph). The star $K_{1,n}$ and the path P_n are examples of trees for all $n \geq 1$. A graph that does not contain any cycles is called a *forest*. That is, a forest is the disjoint union of trees.

The following well-known lemma (e.g., see Theorem 2.20 in Chartrand et al.'s book [15]) is useful for arguing that a given graph contains a tree as a subgraph. Its proof follows from a simple inductive argument on m and is left as an exercise for the reader.

Lemma 1.1 *If T_m is any tree of order $m \geq 2$ and G is a graph that satisfies $\delta(G) \geq m - 1$, then G contains a subgraph that is isomorphic to T_m.*

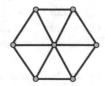 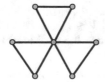

Fig. 1.1 The graphs W_7, F_3, and B_5

The *wheel* W_n is defined by $W_n := K_1 + C_{n-1}$, where $n \geq 4$. Note that some authors prefer to use the notation W_n to mean $K_1 + C_n$, but throughout this book, W_n will be a wheel of order n. For $n \geq 1$, the *fan* F_n is defined by $F_n := K_1 + nK_2$ and has order $2n + 1$. For $n \geq 1$, the *book* B_n is defined by $B_n := K_2 + nK_1$ and has order $n + 2$. Figure 1.1 shows examples of these three types of graph.

In general, this text does not provide proofs of known results on Ramsey numbers (that would easily lead to its own book), The focus instead is on the evaluation of star-critical Ramsey numbers, highlighting the various proof techniques that have been used in the literature, and capturing the current state-of-knowledge on this topic. The asymptotic behavior of star-critical Ramsey numbers does arise in numerous locations. To properly state such results, recall *little-o notation*, where we write $f(n) = o(g(n))$ when f and g are functions of n that satisfy $\lim_{n \to \infty} \frac{f(n)}{g(n)} = 0$.

1.2   Adding or Deleting Edges

Star-critical Ramsey numbers were originally introduced by Hook and Isaak (see [47] and [50]). The reader is encouraged to consult the recent survey article by Hook [49] for a succinct introduction to the topic. Star-critical Ramsey numbers seek to measure the "strength" of a given Ramsey number by determining how many edges must be added between a critical coloring and a vertex in order for the Ramsey property to hold. A special case of this concept had been considered by Erdős et al. in 1978 [27], where they credited the result to Chvátal by personal communication. They proved that it is possible to red/blue color the edges of $K_{r(K_m,K_n)} - e$ so that a red K_m and a blue K_n are avoided. To better frame their result in the context of star-critical Ramsey numbers, we must start with some definitions.

If $1 \leq \ell \leq p$, then the graph $K_p \sqcup K_{1,\ell}$ is defined to consist of the complete graph K_p, a vertex x, and exactly ℓ edges between x and ℓ distinct vertices in the K_p. We note that $K_p \sqcup K_{1,\ell}$ is the same graph as $K_{p+1} - E(K_{1,p-\ell})$. The *star-critical Ramsey number* $r_*(G_1, G_2, \ldots, G_t)$ is the least natural number k such that every t-coloring of $K_{r(G_1,G_2,\ldots,G_t)-1} \sqcup K_{1,k}$ contains a copy of G_i in color i, for some $1 \leq i \leq t$. Two steps are required to prove that $r_*(G_1, G_2, \ldots, G_t) = k$ for a given collection of graphs $G_1, G_2, \ldots, G_t$.

1. Construct a t-coloring of $K_{r(G_1,G_2,...,G_t)-1} \sqcup K_{1,k-1}$ that avoids a monochromatic copy of G_i in color i, for all $1 \leq i \leq t$. From this construction, it follows that $r_*(G_1, G_2, \ldots, G_t) \geq k$ since the addition of $k-1$ edges is not enough to guarantee the Ramsey property.
2. Prove that every t-coloring of $K_{r(G_1,G_2,...,G_t)-1} \sqcup K_{1,k}$ contains a monochromatic copy of G_i in color i, for some $1 \leq i \leq t$. From this step, it follows that $r_*(G_1, G_2, \ldots, G_t) \leq k$ since the addition of k edges is enough to guarantee the Ramsey property.

Since

$$1 \leq r_*(G_1, G_2, \ldots, G_t) \leq r(G_1, G_2, \ldots, G_t) - 1,$$

proving that $r_*(G_1, G_2, \ldots, G_t) = 1$ only requires Step 2 and proving that $r_*(G_1, G_2, \ldots, G_t) = r(G_1, G_2, \ldots, G_t) - 1$ only requires Step 1. We say that $(G_1, G_2, \ldots, G_t)$ is *Ramsey-full* if

$$r_*(G_1, G_2, \ldots, G_t) = r(G_1, G_2, \ldots, G_t) - 1.$$

That is, a collection of graphs is Ramsey-full whenever the removal of a single edge from a complete graph of order $r(G_1, G_2, \ldots, G_t)$ destroys the Ramsey property (the guarantee that there exists a subgraph isomorphic to G_i in color i, for some i). When $G_1 = G_2 = \cdots = G_t$, we just say that G_1 is Ramsey full.

A concept equivalent to star-critical Ramsey numbers was defined in [4], called the deleted edge number. In order to describe it, first define the *k-deleted Ramsey number* $D_k(G_1, G_2, \ldots, G_t)$ to be the minimum natural number p such that every t-coloring of $K_{r(G_1,G_2,...,G_t)} - E(K_{1,k})$ contains a monochromatic copy of G_i in color i, for some $1 \leq i \leq t$. Observe that

$$r(G_1, G_2, \ldots, G_t) \leq D_k(G_1, G_2, \ldots, G_t) \leq r(G_1, G_2, \ldots, G_t) + 1,$$

The *deleted edge number* is defined to be the least natural number k such that

$$D_k(G_1, G_2, \ldots, G_t) = r(G_1, G_2, \ldots, G_t) + 1.$$

The deleted edge number describes how many edges must be removed in order to destroy the Ramsey property, while the star-critical Ramsey number describes the number of edges that must be added to critical colorings to create the Ramsey property. From these definitions, it follows that

$$r(G_1, G_2, \ldots, G_t) = r_*(G_1, G_2, \ldots, G_t) + de(G_1, G_2, \ldots, G_t).$$

Note that it is usually assumed that $t \geq 2$, but the numbers considered so far can be defined for $t = 1$. In this case, we find that $r(G) = |V(G)|$ since $|V(G)|$ vertices are needed to have a monochromatic copy of G and $K_{|V(G)|}$ always contains a subgraph isomorphic to G. One can also confirm that $r_*(G) = \delta(G)$, where $\delta(G)$ is the minimum degree among all vertices in G. It follows that $de(G_1, G_2, \ldots, G_t) = |V(G)| - \delta(G)$.

Next, we state and prove the t-color version of the theorem proved by Erdős et al. [27].

Theorem 1.1 ([27]) *If $n_i \geq 2$ for all $1 \leq i \leq t$, then $(K_{n_1}, K_{n_2}, \ldots, K_{n_t})$ is Ramsey-full. That is,*

$$r_*(K_{n_1}, K_{n_2}, \ldots, K_{n_t}) = r(K_{n_1}, K_{n_2}, \ldots, K_{n_t}) - 1.$$

Proof Let $p = r(K_{n_1}, K_{n_2}, \ldots, K_{n_t})$. This theorem will follow from providing a t-coloring of $K_p - e$ that avoids a copy of K_{n_i} in color i, for all $1 \leq i \leq t$. Begin with a critical graph for $(K_{n_1}, K_{n_2}, \ldots, K_{n_t})$, which is a t-colored K_{p-1} that avoids a copy of K_{n_i} in color i for all $1 \leq i \leq t$. Select a vertex in this graph and label it a. Introduce a new vertex b, adding in edges connecting b to all of the other vertices except for a (we let ab be the missing edge). Color edge bx the same color as edge ax so that the subgraph with a removed is isomorphic to the original critical graph. Since no monochromatic complete graph can include both vertices a and b, we have constructed a t-coloring of $K_p - e$ that avoids a copy of K_{n_i} in color i for all $1 \leq i \leq t$, completing the proof of the theorem. □

In the 2-color case, star-critical Ramsey numbers and deleted edge numbers can be viewed as variations of the *size Ramsey number* $\hat{r}(G_1, G_2)$ introduced by Erdős et al. in 1978 [27]. It is defined to be the minimum number of edges in a graph F such that every red/blue coloring of F results in a red copy of G_1 or a blue copy of G_2. Other related variations include the lower and upper size Ramsey numbers introduced by Erdős and Faudree in 1992 [26]. The *lower size Ramsey number* $\ell(G_1, G_2)$ is the minimum number of edges in any subgraph L of $K_{r(G_1,G_2)}$ such that every red/blue coloring of the edges of L contains a red G_1 or a blue G_2. The *upper size Ramsey number* $u(G_1, G_2)$ is the minimum number such that if a subgraph L of $K_{r(G_1,G_2)}$ has at least $u(G_1, G_2)$ edges, then every red/blue coloring of the edges of L contains a red G_1 or a blue G_2. The inequality

$$\ell(G_1, G_2) \leq \binom{r(G_1, G_2) - 1}{2} + r_*(G_1, G_2) \leq u(G_1, G_2)$$

can be easily confirmed.

1.3 Ramsey-Goodness

Recall that a tree is a connected graph that does not contain any cycles. It is well-known that every tree of order m has size $m - 1$ and every tree contains at least two leaves. It is a straight-forward exercise to prove that every tree can be constructed edge-by-edge, with the resulting graph being a tree at each stage of the construction. This property is particularly useful in proving results about trees using induction.

In 1972, Chvátal and Harary [22] proved the general lower bound

$$r(G_1, G_2) \geq (c(G_1) - 1)(\chi(G_2) - 1) + 1, \qquad (1.1)$$

where $c(G_1)$ is the order of a largest connected component of G_1 (so that $c(G_1) = |V(G_1)|$ whenever G_1 is connected). The construction that led to this result involved replacing all of the vertices in a blue $K_{\chi(G_2)-1}$ with red copies of $K_{c(G_1)-1}$. Chvátal [21] proved in 1977 that this bound is equal to the Ramsey number in the special case where $G_1 = T$ is a tree and $G_2 = K_n$ is a complete graph of order n:

$$r(T, K_n) = (|V(T)| - 1)(n - 1) + 1.$$

In 1983, Burr and Erdős [11] introduced a concept that measured how close a graph is to being a tree. They called a graph G n-good if

$$r(G, K_n) = (c(G) - 1)(n - 1) + 1.$$

Inequality (1.1) was slightly improved by Burr in 1981 [9], resulting in the following theorem. Recall that if $\chi(G_2)$ is the chromatic number of G_2, then $s(G_2)$ is the chromatic surplus, defined to be the minimum number of vertices in any color class among all proper vertex colorings of G_2 that use $\chi(G_2)$ colors.

Theorem 1.2 ([9]) *For all graph G_1 and G_2 satisfying $c(G_1) \geq s(G_2)$,*

$$r(G_1, G_2) \geq (c(G_1) - 1)(\chi(G_2) - 1) + s(G_2).$$

Proof Replace each vertex in a blue $K_{\chi(G_2)-1}$ with copies of red $K_{c(G_1)-1}$-subgraphs. Join to the resulting $K_{(c(G-1)-1)(\chi(G_2)-1)}$ a red $K_{s(G_2)-1}$ using blue edges. The red/blue coloring of $K_{(c(G-1)-1)(\chi(G_2)-1)+s(G_2)-1}$ that has been produced does not contain any red copy of G_1 since the largest red component has order $c(G_1) - 1$. Using all $\chi(G_2)$ colors, every blue subgraph can be properly vertex colored by identifying colors with the complete red subgraph each vertex is contained within. The least order of a color class in such a coloring is $s(G_2) - 1$, preventing G_2 from being a blue subgraph. It follows that

$$r(G_1, G_2) > (c(G_1) - 1)(\chi(G_2) - 1) + s(G_2) - 1,$$

resulting in the statement of the theorem.   □

Burr [9] further generalized the concept of an n-good graph by defining G_1 to be G-good whenever

$$r(G_1, G) = (c(G_1) - 1)(\chi(G) - 1) + s(G).$$

Burr's idea easily extends to a multiset $\mathcal{H} = \{G_1, G_2, \ldots, G_{t-1}\}$ of connected graphs satisfying $r(G_1, G_2, \ldots, G_{t-1}) \geq s(G_t)$:

$$r(G_1, G_2, \ldots, G_t) \geq (r(G_1, G_2, \ldots, G_{t-1}) - 1)(\chi(G_t) - 1) + s(G_t)$$

(e.g., see Theorem 5 of [6]). The multiset $\mathcal{H}$ is G-good if

$$r(G_1, G_2, \ldots, G_{t-1}, G) = (r(G_1, G_2, \ldots, G_{t-1}) - 1)(\chi(G) - 1) + s(G).$$

The advantage of considering G-good multisets (or G-good graphs) is that there is a natural critical coloring upon which one can construct lower bounds for the corresponding star-critical Ramsey numbers. The next three theorems offer such lower bounds. The following theorem is Theorem 2.1 in [5].

Theorem 1.3 ([5]) *Let $G_1, G_2, \ldots, G_{t-1}$ be connected graphs, each having order at least two. If G_t is a graph such that $\{G_1, G_2, \ldots, G_{t-1}\}$ is a G_t-good multiset satisfying $r(G_1, G_2, \ldots, G_{t-1}) \geq s(G_t)$, then*

$$r_*(G_1, G_2, \ldots, G_{t-1}, G_t) \geq r_*(G_1, G_2, \ldots, G_{t-1}) + r(G_1, G_2, \ldots, G_{t-1}, G_t)$$
$$- r(G_1, G_2, \ldots, G_{t-1}).$$

Proof This theorem is most easily proved from the perspective of deleted edge numbers. Let $m = r(G_1, G_2, \ldots, G_{t-1})$. We will construct a t-coloring of

$$K_{(m-1)(\chi(G_t)-1)+s(G_t)} - E(K_{1,de(G_1,G_2,\ldots,G_{t-1})})$$

that lacks a monochromatic copy of G_i in color i, for all $1 \leq i \leq t$. Start with a $(t-1)$-coloring of $K_m - E(K_{1,de(G_1,G_2,\ldots,G_{t-1})})$ that lacks a monochromatic copy of G_i in color i, for all $1 \leq i \leq t-1$. Call this graph A_1 and let v be the center vertex for the missing star. Let $A_2, A_3, \ldots, A_{\chi(G_t)-1}$ be copies of $A_1 - \{v\}$ so that each one is a $(t-1)$-coloring of K_{m-1} that lacks a monochromatic copy of G_i in color i, for all $1 \leq i \leq t-1$. Let $A_{\chi(G_t)}$ be formed by taking another copy of $A_1 - \{v\}$ and removing $m - s(G_t)$ vertices. That is, $A_{\chi(G_t)}$ is a $(t-1)$-coloring of $K_{s(G_t)-1}$ that lacks a monochromatic copy of G_i in color i, for all $1 \leq i \leq t-1$. Consider the union

$$\bigcup_{1 \leq j \leq \chi(G_t)} A_j,$$

and color all of the edges interconnecting the different A_j in color t. The resulting

$$K_{(m-1)(\chi(G_t)-1)+s(G_t)} - E(K_{1,de(G_1,G_2,\ldots,G_{t-1})})$$

lacks a monochromatic copy of G_i in color i, for all $1 \leq i \leq t-1$.

It remains to be shown that it also lacks a copy of G_t in color t. If $s(G_t) = 1$, then any subgraph in color t can be properly vertex colored by assigning colors based on the vertex sets A_j. Since only $\chi(G_t) - 1$ colors are needed, no subgraph isomorphic to G_t exists in color t. If $s(G_t) > 1$, then coloring the vertices according to the A_j they are contained in produces a proper vertex coloring of any subgraph in color t using $chi(G_t)$ colors. Since some color class contains only $s(G_t)$ colors, no copy of G_t exists in color t. It follows that

$$de(G_1, G_2, \ldots, G_{t-1}, G_t) \leq de(G_1, G_2, \ldots, G_{t-1}),$$

from which the theorem follows. $\square$

If G_1 is a connected graph such that $|V(G_1)| \geq s(G_2)$ and G_1 is G_2-good, then $r_*(G_1) = \delta(G_1), r(G_1) = |V(G_1)|$, and Theorem 1.3 simplifies to

$$r_*(G_1, G_2) \geq r(G_1, G_2) + \delta(G_1) - |V(G_1)|$$
$$\geq (|V(G_1)| - 1)(\chi(G_2) - 1) + \delta(G_1) - (|V(G_1)| - 1) - 1$$
$$\geq (|V(G_1)| - 1)(\chi(G_2) - 2) + s(G_2) + \delta(G_1) - 1.$$

Consider a proper vertex coloring of G_2 using $\chi(G_2)$ colors in which there exists a color class having $s(G_2)$ vertices. Let $V_1, V_2, \ldots, V_{\chi(G_2)}$ be the distinct color classes of G_2 such that $|V_1| \leq |V_2| \leq \cdots \leq |V_{\chi(G_2)}|$ and $|V_1| = s(G_2)$. For $x \in V_1$, define $\deg_{V_i}(x)$ to be the number of edges joining x to V_i, for $2 \leq i \leq \chi(G_t)$. Define

$$\tau(G_2) := \min\{\deg_{V_i}(x) \mid x \in V_1, \ 2 \leq i \leq \chi(G_2)\},$$

which is the minimum degree among all vertices in V_1 to some V_i, where $2 \leq i \leq \chi(G_2)$. The following lower bound was proved by Zhang et al. [95] in 2016.

Theorem 1.4 ([95]) *Let G_1 be a connected graph of order at least two that is G_2-good. If $s(G_2) = 1$, $\delta(G_1) = 1$, or $\kappa(G_1) \geq 2$, then*

$$r_*(G_1, G_2) \geq (|V(G_1)| - 1)(\chi(G_2) - 2) + s(G_2) + \delta(G_1) + \tau(G_2) - 2.$$

Proof If $\chi(G_2) = 1$, then the inequality holds trivially. So assume that $\chi(G_2) \geq 2$. If $|V(G_1)| \leq s(G_2)$, then

$$(|V(G_1)| - 1)(\chi(G_2) - 1) + s(G_2) \leq (s(G_2) - 1)(\chi(G_2) - 1) + s(G_2)$$
$$\leq s(G_2)\chi(G_2) - \chi(G_2) + 1 < |V(G_2)|.$$

However, $r(G_1, G_2) \geq |V(G_2)|$, from which it follows that G_1 is not G_2-good. So, assume that $|V(G_1)| \geq s(G_2) + 1$. Also, note that $\tau(G_2) \geq 1$, otherwise there exists a vertex $x \in V_1$ such that there is some i such that $2 \leq i \leq \chi(G_2)$ and no edge joins

x with V_i. Then x can be given the same color as the vertices in V_i, from which it follows that either $\chi(G_2) \leq \chi(G_2) - 1$ (which is not possible) or $s(G_2) \leq |V_1| - 1$, contradicting the choice of V_1. It follows that $\tau(G_2) \geq 1$.

Since G_1 is G_2-good,

$$r(G_1, G_2) = (|V(G_1)| - 1)(\chi(G_2) - 1) + s(G_2).$$

Consider the red/blue coloring of $K_{(|V(G_1)|-1)(\chi(G_2)-1)+s(G_2)-1}$ formed by replacing $\chi(G_2) - 1$ of the vertices in a blue $K_{\chi(G_2)}$ with red copies of $K_{|V(G_1)|-1}$ and one vertex with a red copy of $K_{s(G_2)-1}$. Let $X_1, X_2, \ldots, X_{\chi(G_2)-1}$ denote the vertex sets of the red $K_{|V(G_1)|-1}$-subgraphs and let $X_{\chi(G_2)}$ denote the vertex set of the red $K_{s(G_2)-1}$. Introduce a vertex v and join v to all of the vertices in $X_2 \cup X_3 \cup \cdots \cup X_{\chi(G_2)-1}$ with blue edges. and to the vertices in $X_{\chi(G_2)}$ with red edges. Join v to $\delta(G_1) - 1$ vertices in X_1 with red edges and $\min\{\tau(G_2) - 1, |V(G_1)| - \delta(G_1)\}$ vertices in X_1 with blue edges. If $\tau(G_2) - 1 \geq |V(G_1)| - \delta(G_1)$, then the coloring corresponds with a red/blue coloring of $K_{r(G_1,G_2)}$. Otherwise, $\tau(G_2) - 1 < |V(G_1)| - \delta(G_1)$, resulting in a red/blue coloring of $K_{r(G_1,G_2)-1} \sqcup K_{1,m}$, where

$$m = (V(G_1)| - 1)(\chi(G_2) - 2) + s(G_2) + \delta(G_1) + \tau(G_2) - 3.$$

To see that this coloring contains no red copy of G_1, first consider the case $\delta(G_1) = 1$. Then every red component has at most $|V(G_1)| - 1$ vertices. If $\delta(G_1) \geq 2$, then the only red component containing at least $|V(G_1)|$ vertices is spanned by the red edges in $X_1 \cup X_{\chi(G_2)} \cup \{v\}$. If $s(G_2) = 1$, then $X_{\chi(G_2)} = \emptyset$ and this red component has order $|V(G_1)|$. As it has a vertex of order $\delta(G_1) - 1$, it does not contains a red copy of G_1. If $s(G_2) \geq 2$, then it is assumed that $\kappa(G_1) = 1$. Since the red subgraphs induced by $X_1 \cup \{v\}$ and $X_{\chi(G_2)} \cup \{v\}$ do not contain red G_1-subgraphs, if a red G_1 is contained in $X_1 \cup X_{\chi(G_2)} \cup \{v\}$, then it must contain at least one vertex in X_1 and at least one vertex in $X_{\chi(G_2)}$. It follows that $\kappa(G_1) \leq 1$, contradicting the assumption that $\kappa(G_1) \geq 2$.

Denote the subgraph spanned by blue edges by G_B and suppose that G_B contains a blue copy of G_2. If $s(G_2) = 1$ and $\tau(G_2) = 1$, then $X_{\chi(G_2)} = \emptyset$ and $\chi(G_B) = \chi(G_2) - 1$. It follows that G_B does not contain a copy of G_2. If $s(G_2) = 1$ and $\tau(G_2) \geq 2$, or if $s(G_2) \geq 2$, then $\chi(G_B) = \chi(G_2)$ and any proper vertex coloring of G_B restricted to $V(G_2)$ is also a proper vertex coloring of G_2. For $1 \leq i \leq \chi(G_2)$, color the vertices in X_i with color i and color v using color $\chi(G_2)$. Then $\chi(G_B) = \chi(G_2)$, $s(G_B) = s(G_2)$, and $s(G_2)$ vertices have color $\chi(G_2)$, and v is adjacent t at most $\tau(G_2) - 1$ vertices in color class X_1. When restricting this proper vertex coloring to $V(G_2)$, it follows that $\tau(G_1) \leq \tau(G_2) - 1$, leading to a contradiction. Thus, G_B does not contain G_2 as a subgraph.

Finally, if $\tau(G_2) - 1 \geq |V(G_1)| - \delta(G_1)$, then the coloring corresponds with a coloring of $K_{r(G_1,G_2)}$ that lacks a red G_1 and a blue G_2, which is a contradiction. Thus, $\tau(G_2) - 1 < |V(G_1)| - \delta(G_1)$, and the coloring corresponds with $K_{r(G_1,G_2)-1} \sqcup K_{1,m}$, where

$$m = (|V(G_1)| - 1)(\chi(G_2) - 2) + s(G_2) + \delta(G_1) + \tau(G_2) - 3,$$

and the theorem follows.   □

Hao and Lin (see [42] and [43]) offered the following variation of the previous theorem.

Theorem 1.5 ([42, 43]) *Let G_2 be a graph with $\chi(G_2) \geq 2$ and let G_1 be a connected graph that is G_2-good with order $|V(G_1)| \geq s(G_2) + 1$. Then*

$$r_*(G_1, G_2) \geq (|V(G_1)| - 1)(\chi(G_2) - 2) + \min\{|V(G_1)|, \delta(G_1) + \tau(G_2) - 1\}.$$

If G_1 does not contain any cut vertices or if $\delta(G_1) = 1$, then

$$r_*(G_1, G_2) \geq (|V(G_1)| - 1)(\chi(G_2) - 2) + \min\{|V(G_1)|, \delta(G_1) + \tau(G_2) - 1\}$$
$$+ s(G_2) - 1.$$

Proof Let G_1 be G_2-good, from which it follows that

$$r(G_1, G_2) = (|V(G_1)| - 1)(\chi(G_2) - 1) + s(G_2).$$

Consider the red/blue coloring of $K_{(|V(G_1)|-1)(\chi(G_2)-1)+s(G_2)-1}$ formed by replacing $\chi(G_2) - 1$ of the vertices in a blue $K_{\chi(G_2)}$ with red copies of $K_{|V(G_1)|-1}$ and one vertex with a red copy of $K_{s(G_2)-1}$. Let $X_1, X_2, \ldots, X_{\chi(G_2)-1}$ denote the vertex sets of the red $K_{|V(G_1)|-1}$-subgraphs and let $X_{\chi(G_2)}$ denote the vertex set of the red $K_{s(G_2)-1}$. Select a subset $D \subseteq X_{\chi(G_2)-1}$ of cardinality $\delta(G_1) - 1$ and a subset $T \subseteq X_{\chi(G_2)-1} - D$ of cardinality $\min\{|V(G_1)|, \delta(G_1) + \tau(G_2) - 1\}$.

Introduce a new vertex v, joining it to D with red edges and to $X_1 \cup X_2 \cup \cdots \cup X_{\chi(G_2)-2} \cup T$ with blue edges. Denote the subgraph spanned by the red edges by G_R and the subgraph spanned by the blue edges by G_B. Then G_B does not contain a subgraph isomorphic to G_2 since $\tau(G_B) \leq \tau(G_2) - 1$. Also, any subgraph of G_R of order $|V(G_1)|$ has minimum degree at most $\delta(G_1) - 1$, preventing a red G_1 from being a subgraph of G_R. From this construction, it follows that

$$r_*(G_1, G_2) \geq (|V(G_1)| - 1)(\chi(G_2) - 2) + (\delta(G_1) - 1)$$
$$+ \min\{\tau(G_2) - 1, |V(G_1)| - \delta(G_1)\} + 1$$
$$\geq (|V(G_1)| - 1)(\chi(G_2) - 1) + \min\{|V(G_1)|, \delta(G_1) + \tau(G_2) - 1\},$$

completing the proof of first statement of the theorem.

Now assume that G_1 does not contain any cut vertices and define G'_R to be red graph the obtained from G_R by adding in all edges joining v and $D \cup V_{\chi(G_2)}$. Note that G'_R does not contain a copy of G_1 as a subgraph, otherwise, there exists a subgraph F such that $G_1 \subseteq F \subseteq G'_R$. Since both $X_{\chi(G_2)-1} \cup \{v\}$ and $X_{\chi(G_2)} \cup \{v\}$ contain no G_1, it follows that $v \in V(F)$, $D \cap V(F) \neq \emptyset$, and $V_{\chi(G_2)} \cap V(F) \neq \emptyset$. This is a contradiction since v is a cut vertex for F. If $\delta(G_1) = 1$, then the assertion follows by forming G''_R from G_R by adding in all edges joining v and $V_{\chi(G_2)}$.   □

Chapter 2
Star-Critical Ramsey Numbers Involving Various Graphs

2.1 Trees Versus Trees

In this section, star-critical Ramsey numbers are considered where all of the arguments are trees. The earliest result in this area was due to Hook (see [47] and [48]) and concerns the case of a path versus a path.

2.1.1 Paths Versus Paths

In 1967, Gerencsér and Gyárfás [34] proved that if $n \geq m \geq 2$, then

$$r(P_m, P_n) = \left\lfloor \frac{m}{2} \right\rfloor + n - 1.$$

In [47] and [48], Hook determined the value of $r_*(P_m, P_n)$ after completely classifying the critical colorings of (P_m, P_n). When $n \geq m \geq 4$, the critical colorings of (P_m, P_n) are shown in Fig. 2.1. Suppose that A_k is any red/blue coloring of K_k. Let G_1 be the coloring formed by replacing one vertex in a red K_2 with a blue K_{n-1} and the other vertex with $A_{\lfloor \frac{m}{2} \rfloor - 1}$. The coloring G_2 is given by replacing one of the vertices in a red K_2 with a blue $K_{n-1} - e$ (the missing edge is colored red) and the other vertex with $A_{\lfloor \frac{m}{2} \rfloor - 1}$. The coloring G_3 is formed by replacing one of the vertices in a blue K_2 with a red K_{n-1} and the other vertex with a $A_{\lfloor \frac{m}{2} \rfloor - 1}$. The coloring G_4 is given by replacing one of the vertices in a blue K_2 with a red $K_{n-1} - e$ (the missing edge is colored blue) and the other vertex with $A_{\lfloor \frac{m}{2} \rfloor - 1}$. The coloring G_5 is given by replacing one of the vertices in a blue K_2 with a red K_{m-1} and the other vertex with $A_{\frac{n}{2} - 1}$.

Theorem 2.1 ([47, 48]) *If $n \geq m \geq 4$, then every critical coloring of (P_m, P_n) corresponds with a coloring described in Fig. 2.1. The colorings given by G_1 are*

© The Author(s), under exclusive license to Springer Nature Switzerland AG 2023
M. R. Budden, *Star-Critical Ramsey Numbers for Graphs*, SpringerBriefs in
Mathematics, https://doi.org/10.1007/978-3-031-29981-0_2

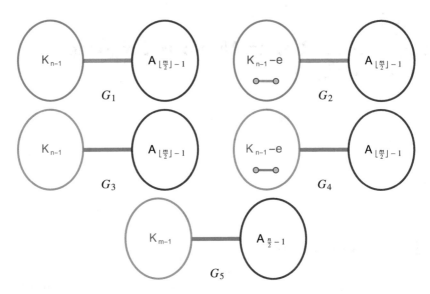

Fig. 2.1 Critical colorings for (P_m, P_n), when $n \geq m \geq 4$

critical colorings for (P_m, P_n) for all $n \geq m$. The colorings given by G_2 and G_4 are critical colorings for (P_m, P_n) only when m is odd. The colorings given by G_3 and G_4 are critical colorings for (P_m, P_n) when $m = n$. The colorings given by G_5 are critical colorings for (P_m, P_n) when $n = m + 1$ and m is odd.

Theorem 2.2 ([47, 48]) *If $n \geq m \geq 2$, then $r_*(P_m, P_n) = \lceil \frac{m}{2} \rceil$.*

Proof When $m = 2$, $P_2 = K_2$ and $r(P_2, P_n) = n$ for all $n \geq 2$. Since a critical coloring for (P_2, P_n) cannot have any red edges, it must be a blue K_{n-1}. Introduce a vertex v and a single edge joining v to the blue K_{n-1}. If the edge is red, a red P_2 is formed. Otherwise, the edge is blue and a blue P_n is formed with v being one of the leaves of the P_n. It follows that $r_*(P_2, P_n) = 1$ for all $n \geq 2$.

If $m = 3$, then $r(P_3, P_n) = n$ for all $n \geq 3$. Consider a red K_2 in which one of the vertices is replaced with a blue K_{n-1}. The resulting $K_{n-1} \sqcup K_{1,1}$ avoids a red P_3 and a blue P_n, from which it follows that $r_*(P_3, P_n) \geq 2$, for all $n \geq 3$. To prove the reverse inequality, consider a red/blue coloring of $K_{n-1} \sqcup K_{1,2}$ and let v be the center vertex for the missing star. Removing v results in a red/blue coloring of K_{n-1}. If this coloring avoids a red P_3 and a blue P_n, the the subgraph spanned by the red edges must be a matching or an empty graph. The subgraph spanned by the blue edges necessarily contains a P_{n-1} with any selected vertex as a leaf. If either of the two edges incident with v are blue, then a blue P_n is formed. Otherwise, they are both red and form a red P_3. It follows that $r_*(P_3, P_n) \geq 2$, for all $n \geq 3$.

When $n \geq m \geq 4$, begin by replacing one vertex in a red K_2 with a blue K_{n-1} and the other vertex with a blue $K_{\lfloor \frac{m}{2} \rfloor - 1}$. Add in a vertex v and join it to the blue $K_{\lfloor \frac{m}{2} \rfloor - 1}$ via blue edges. If m is odd, also join v to a single vertex in the blue K_{n-1}

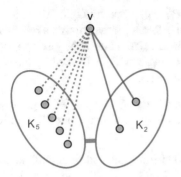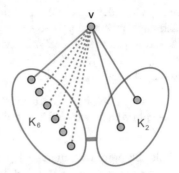

Fig. 2.2 Red/blue colorings of $K_{n+\lfloor\frac{m}{2}\rfloor-2} \sqcup K_{1,\lceil\frac{m}{2}\rceil-1}$ that avoid a red P_m and a blue P_n. The first image is the case $m = n = 6$ and the second image is the case $m = n = 7$

using a red edge. The resulting $K_{n+\lfloor\frac{m}{2}\rfloor-2} \sqcup K_{1,\lceil\frac{m}{2}\rceil-1}$ avoids a red P_m and a blue P_n, from which it follows that $r_*(P_m, P_n) \geq \lceil\frac{m}{2}\rceil$ (Fig. 2.2).

To prove the reverse inequality, consider a red/blue coloring of $K_{n+\lfloor\frac{m}{2}\rfloor-2} \sqcup K_{1,\lceil\frac{m}{2}\rceil}$ and let v be the center vertex for the missing star. Removing v results in a red/blue coloring of $K_{n+\lfloor\frac{m}{2}\rfloor-2}$ that we assume avoids a red P_m and a blue P_n. So, it must be one of the colorings described in Theorem 2.1. We divide the rest of the proof into cases, based on each of the classes of critical colorings.

Case 1   Consider the critical colorings of (P_m, P_n) given by G_1 in Theorem 2.1. If m is even, then v must join to the blue K_{n-1} with at least one edge. If the edge is blue, then a blue P_n is formed. If the edge is red, then it can be joined onto a leaf in a red P_{m-1} that is formed by the edges joining the blue K_{n-1} and $A_{\frac{m}{2}-1}$, resulting in a red P_m. If m is odd, then v joins to the blue K_{n-1} via at least two edges. If any such edge is blue, then a blue P_n can be formed. Assume that both such edges are red. Then the red P_3 they form can be joined to a leaf in a red P_{m-2} formed by the edges joining the blue K_{n-1} and the $A_{\frac{m-1}{2}-1}$, resulting in a red P_m. In both cases, $\lceil\frac{m}{2}\rceil$ edges force the existence of a red P_m or a blue P_n.

Case 2   Consider the critical colorings of (P_m, P_n) given by G_2 in Theorem 2.1. This case only occurs when m is odd, which means v joins to at least two edges in the blue $K_{n-1} - e$. If any such edge is blue, then a blue P_n is formed. So, assume that two such red edges exist. Then the red P_3 they form can be joined to a leaf in a red P_{m-2} formed by the edges joining the blue K_{n-1} and the $A_{\frac{m-1}{2}-1}$, resulting in a red P_m.

Case 3   Consider the critical colorings of (P_m, P_n) given by G_3 in Theorem 2.1. This case only occurs when $m = n$, and is proved following the same logic as in Case 1, but with the colors red and blue interchanged.

Case 4   Consider the critical colorings of (P_m, P_n) given by G_4 in Theorem 2.1. This case only occurs when $m = n$ is odd. The proof follows the same logic as in Case 2, but with the colors red and blue interchanged.

Case 5 Consider the critical colorings of (P_m, P_n) given by G_5 in Theorem 2.1. This case only occurs when $n = m + 1$ and m is odd. If there exists a red edge to the red K_{m-1}, then a red P_m is formed. If there exists a blue edge joining v to the red K_{m-1}, then joining this edge onto a leaf in a red P_{n-1} formed by the edges joining the red K_{m-1} to the $A_{\frac{n}{2}-1}$ results in a blue P_n.

In all five cases, joining v to a critical coloring of (P_m, P_n) using $\lceil \frac{m}{2} \rceil$ edges results in a graph in which every red/blue coloring contains a red P_m or a blue P_n. It follows that $r_*(P_m, P_n) \leq \lceil \frac{m}{2} \rceil$.   □

While the proof of Lemma 2.1 was excluded from this text, it should be noted that the dependence of the proof of Theorem 2.2 on Lemma 2.1 can be avoided for most cases by using the path-cycle Ramsey number proved by Faudree et al. in 1974 [31]. For $n \geq m \geq 2$, they proved that

$$r(P_m, C_n) = \begin{cases} n + \lfloor \frac{m}{2} \rfloor - 1 & \text{if } n \text{ is even} \\ \max\{n + \lfloor \frac{m}{2} \rfloor - 1, 2m - 1\} & \text{if } n \text{ is odd.} \end{cases}$$

When $n > m \geq 4$, where n is odd or $n \geq 2m - \lfloor \frac{m}{2} \rfloor + 1$, it follows that

$$r(P_m, C_{n-1}) = n + \lfloor \frac{m}{2} \rfloor - 2.$$

Consider a red/blue coloring of $K_{n+\lfloor \frac{m}{2} \rfloor - 2} \sqcup K_{1, \lceil \frac{m}{2} \rceil}$ and let v be the center vertex for the missing star. Removing v results in a red/blue coloring of $K_{n+\lfloor \frac{m}{2} \rfloor - 2}$ that we assume avoids a red P_m and a blue P_n. Hence, it must contain a blue C_{n-1}, which we denote by $x_1 x_2 \cdots x_{n-1} x_1$. Denote the other vertices in the red/blue $K_{n+\lfloor \frac{m}{2} \rfloor - 2}$ by $y_1, y_2, \ldots, y_{\lfloor \frac{m}{2} \rfloor - 1}$. Let $X = \{x_1, x_2, \ldots, x_{n-1}\}$ and $Y = \{y_1, y_2, \ldots, y_{\lfloor \frac{m}{2} \rfloor - 1}\}$. If any $x_i y_j$ is blue, then $y_j x_i x_{i+1} \cdots x_{n-1} x_1 x_2 \cdots x_{i-1}$ is a blue P_n. So, all edges joining X and Y must be red.

Suppose that m is odd. Then $\lfloor \frac{m}{2} \rfloor + 1 = \lceil \frac{m}{2} \rceil$ and it follows that v must be incident with at least two of the vertices in X. If either of those edges is blue, say vx_i, then $vx_i x_{i+1} \cdots x_{n-1} x_1 x_2 \cdots x_{i-1}$ is a blue P_n. So, assume both of the edges are red, say vx_i and vx_j ($i \neq j$). Then there exists a red P_{m-2} formed by the edges joining X and Y that has x_i as one of its endpoints and that does not include x_j. This path, along with the red edges $x_i v$ and vx_j, form a red P_m.

Suppose that m is even. Then $\lfloor \frac{m}{2} \rfloor = \frac{m}{2} = \lceil \frac{m}{2} \rceil$ and v must be incident with at least one vertex in X. If some vx_i is blue, then $vx_i x_{i+1} \cdots x_{n-1} x_1 x_2 \cdots x_{i-1}$ is a blue P_n. If vx_i is red, then joining it to the end of a red path P_{m-1} that exists between the sets X and Y results in a red P_m. It follows that $r_*(P_m, P_n) \leq \lceil \frac{m}{2} \rceil$ for $n > m \geq 4$, when n is odd or $n \geq 2m - \lfloor \frac{m}{2} \rfloor + 1$.

For multicolor star-critical Ramsey numbers involving paths, the following is Theorem 3.6 of [5].

Theorem 2.3 ([5]) *If $t \geq 1$ is odd, then $r_*^t(P_3) = 1$.*

Proof It was shown in [51] that when $t \geq 1$ is odd, $r^t(P_3) = t + 2$. Consider a t-coloring of K_{t+1} that lacks a monochromatic P_3. Such a coloring has every vertex incident with exactly one edge in each of the t colors. Add in a new vertex and connect it with a single edge to some vertex in the K_{t+1}. Regardless of what color is assigned to this edge, a monochromatic P_3 is produced. □

2.1.2 Stars Versus Stars

Now, we turn our attention to star-critical Ramsey numbers involving stars. In [45], it was proved that if $m, n \geq 1$, then

$$r(K_{1,m}, K_{1,n}) = \begin{cases} m + n & \text{if } m \text{ or } n \text{ are odd} \\ m + n - 1 & \text{if } m \text{ and } n \text{ are both even.} \end{cases}$$

This result was extended to more than two colors by Burr and Roberts [14] in 1973. To describe their result, let $S = \{m_1, m_2, \dots, m_t\}$, with each $m_i \geq 2$, and define

$$N = \sum_{i=1}^{t} m_i.$$

Letting k denote the number of elements in S that are even, Burr and Roberts proved that

$$r(K_{1,m_1}, K_{1,m_2}, \dots, K_{1,m_t}) = \begin{cases} N - t + 1 & \text{if } k \geq 2 \text{ is even} \\ N - t + 2 & \text{otherwise.} \end{cases} \tag{2.1}$$

In 1978, Erdős, Fauree, Rousseau, and Schelp (see Theorem 2 of [27]) observed that every 2-coloring of the edges of $K_{1,m+n-1}$ contains a copy of $K_{1,m}$ in color 1 or a copy of $K_{1,n}$ in color 2. Generalizing this observation to t colors leads to the following theorem.

Theorem 2.4 ([4, 5]) *Let $S = \{m_1, m_2, \dots, m_t\}$ with each $m_i \geq 2$ and let k be the number of elements in S that are even. If $k = 0$ or k is odd, then*

$$r_*(K_{1,m_1}, K_{1,m_2}, \dots, K_{1,m_t}) = 1.$$

Proof When $k = 0$ or k is odd, Eq. (2.1) implies that

$$r(K_{1,m_1}, K_{1,m_2}, \dots, K_{1,m_t}) = N - t + 2.$$

Removing $N - t$ edges incident with a fixed vertex in a t-colored K_{N-t+2} still leaves one vertex having degree $N - t + 1$. By the pigeonhole principle, this vertex must be incident with at least m_i edges in color i, for some $1 \leq i \leq t$. It follows that every

t-coloring of $K_{N-t+1} \sqcup K_{1,1}$ contains a monochromatic K_{1,m_i} in color i, for some $1 \leq i \leq t$, from which the theorem follows.   □

Erdős and Faudree [26] first observed that when $m, n \geq 2$ are both even, $(K_{1,m}, K_{1,n})$ is Ramsey-full. This result was then generalized to t colors in the following theorem from [5]. In order to prove this theorem, we need to define the *Ramsey multiplicity* $R(G_1, G_2, \ldots, G_t)$, first introduced in 1974 by Harary and Prins [46], to be the smallest possible total number of G_1 in color 1, G_2 in color 2, $\ldots$, G_t in color t, among all t-colorings of $K_{r(G_1, G_2, \ldots, G_t)}$. In the case of stars, Jacobson [52] proved that

$$R(K_{1,m_1}, K_{1,m_2}, \ldots, K_{1,m_t}) = \begin{cases} k/2 & \text{if } k \geq 2 \text{ is even} \\ N - t + 2 & \text{otherwise.} \end{cases} \tag{2.2}$$

Theorem 2.5 ([4, 5]) *If $m_1, m_2, \ldots, m_t$ are integers greater than 1, exactly two of which are even, then*

$$r_*(K_{1,m_1}, K_{1,m_2}, \ldots, K_{1,m_t}) = N - t.$$

Proof Assuming that $k = 2$, Eq. (2.2) implies that there exists a t-coloring of $K_{r(K_{1,m_1}, K_{1,m_2}, \ldots, K_{1,m_t})}$ that contains a single monochromatic K_{1,m_i} in some color i, and which does not contains a K_{1,m_i+1} in color i. It follows that a single edge in color i can be removed to produce a $K_{r(G_1, G_2, \ldots, G_t)} - e$ that lacks a monochromatic copy of K_{1,m_i} in color i, for all $1 \leq i \leq t$. This observation, along with Eq. (2.1), implies the statement of the theorem.   □

At the present time, the evaluation of $r_*(K_{1,m_1}, K_{1,m_2}, \ldots, K_{1,m_t})$ when $k > 2$ is even is unknown. In [5], it was conjectured that the star-critical Ramsey number is equal to $N - t$ for these remaining cases (see Open Problem 1 in Sect. 4.1).

Conjecture 2.1 ([5]) *If $m_1, m_2, \ldots, m_t$ are integers greater than 1, exactly k of which are even, then*

$$r_*(K_{1,m_1}, K_{1,m_2}, \ldots, K_{1,m_t}) = \begin{cases} N - t & \text{if } k \geq 2 \text{ is even} \\ 1 & \text{otherwise.} \end{cases}$$

2.1.3   Other Trees

The following lemma is similar to Lemma 1.1, but focuses on trees that are not stars. It is a special case of a lemma proved by Guo and Volkmann (see Lemma 2.3 of [38]). Its proof is left as an exercise for the reader. Note that among all trees of order m, the star $K_{1,m-1}$ is the only tree that has a vertex of order $m - 1$. All non-star trees of order m have maximum degree at most $m - 2$.

Lemma 2.1 ([38]) *If T_m is any non-star tree of order $m \geq 4$ and G is any connected graph of order $|V(G)| \geq m$ satisfying $\delta(G) \geq m - 2$, then G contains a subgraph isomorphic to T_m.*

In 1974, Burr [8] proved that if T_m is any tree of order m for which $(m-1)|(n-1)$, then

$$r(T_m, K_{1,n}) = m + n - 1.$$

In the following theorem, we show how the previous lemma can be used to prove an upper bound for $r_*(T_m, K_{1,n})$ when T_m is a non-star tree in which m satisfies a certain divisibility property.

Theorem 2.6 ([4]) *Let T_m be a non-star tree of order $m \geq 4$. Then for all $n \geq 2$ such that $(m-1)|(n-1)$,*

$$r_*(T_m, K_{1,n}) \leq m + n - 3.$$

Proof Consider a red/blue coloring of $K_{m+n-1} - e$ and denote by a and b the two vertices of degree $m + n - 3$. Denote the subgraph spanned by the red edges by G_R and the subgraph spanned by the blue edges by G_B. If a blue $K_{1,n}$ is avoided, then $\Delta(G_B) \leq n - 1$. It follows that

$$\delta(G_R) \geq m + n - 3 - (n - 1) = m - 2.$$

In order to apply Lemma 2.1, we must argue that G_R contains a connected component having order at least m. Let G be a largest connected component of G_R. Then G must contain some vertex v other than a or b, otherwise G would be an empty graph. Such an x has degree

$$\deg_G(x) \geq m + n - 2 - (n - 1) = m - 1,$$

forcing G to have order at least m. Since all of the hypotheses of Lemma 2.1 are satisfied, it follows that G (and hence, G_R) contains a subgraph that is isomorphic to T_m. $\square$

Parsons [75] proved that for all $m, n \in \mathbb{N}$ satisfying $m \geq 2n - 1$,

$$r(P_m, K_{1,n}) = m.$$

In 2021, Wang et al. [88] considered the corresponding star-critical Ramsey number. Their proof requires the following well-known lemma concerning vertex degrees and the Hamiltonicity of graphs.

Lemma 2.2 ([20, 25]) *Let G be a graph of order $n \geq 3$ that satisfies at least one of the following statements.*

1. *The minimum degree of G satisfies $\delta(G) \geq \lceil \frac{n}{2} \rceil$.*
2. *For all $i < \frac{n}{2}$, either $d_i \geq i + 1$ or $d_{n-i} \geq n - i$, where $d_1 \leq d_2 \leq \cdots \leq d_n$ is the degree sequence for G.*

Then G is Hamiltonian.

Theorem 2.7 ([88]) *For all $m, n \in \mathbb{N}$ that satisfy $m \geq 2n + 1$, $r_*(P_m, K_{1,n}) = n$.*

Proof For the lower bound, begin with a red K_{m-1}, add in a vertex v, and join v to $n - 1$ vertices in the K_{m-1} using blue edges. Other edges between v and the K_{m-1} are missing. The resulting red/blue coloring of $K_{m-1} \sqcup K_{1,n-1}$ avoids a red P_m and a blue $K_{1,n}$. It follows that $r_*(P_m., K_{1,n}) \geq n$.

To prove the reverse inequality, consider a red/blue coloring of $K_{m-1} \sqcup K_{1,n}$ and let v be the center vertex of the missing star. Removing v produces a red/blue K_{m-1}. If a blue $K_{1,n}$ is avoided, then each vertex has red degree at least $m - 2 - (n - 1) = m - n - 1 \geq \lceil \frac{m}{2} \rceil$. By Lemma 2.2, there must exist a red C_{m-1}. Denote the vertices in this cycle by $x_1 x_2 \cdots x_{m-1} x_1$. If all n of the edges incident with v are blue, then a blue $K_{1,n}$ is formed. Otherwise, some edge incident with v must be red, say vx_1, and $vx_1 x_2 \cdots x_{m-1}$ forms a red P_m. Thus, $r_*(P_m, K_{1,n}) \leq n$.   □

In the next theorem, we consider trees of order $m \geq 4$ that have maximum degree equal to $m - 2$. Such a tree is necessarily a broom in that it can be formed by adding an edge between a vertex in a P_2 and the center vertex in $K_{1,m-3}$. We denote by T_m^* the unique tree of order $m \geq 4$ satisfying $\Delta(T_m^*) = m - 2$. In [38], Guo and Volkmann proved that if $m, n \geq 5$ are integers such that either $(m - 1)|(n - 3)$ or $(n - 1)|(m - 3)$, then

$$r(T_m^*, T_n^*) = m + n - 3.$$

The following theorem offers an upper bound for the corresponding star-critical Ramsey number and was proved in Corollary 3.7 of [4]. It is unlikely that this upper bound is the actual value and the reader is encouraged to consider ways in which this upper bound may be improved (see Open Problem 2 in Sect. 4.1).

Theorem 2.8 ([4]) *Let T_m^* and T_n^* be trees of orders $m \geq 5$ and $n \geq 5$ satisfying*

$$\Delta(T_m^*) = m - 2 \quad and \quad \Delta(T_n^*) = n - 2.$$

If $(m - 1)|(n - 3)$ or $(n - 1)|(m - 3)$, then $r_*(T_m^*, T_n^*) \leq m + n - 5$.

Proof By symmetry, it is sufficient to prove the theorem for only one of the given divisibility properties. Without loss of generality, assume that $(m - 1)t = (n - 3)$ for some natural number t. Consider a red/blue coloring of

$$K_{m+n-4} \sqcup K_{1,m+n-5} = K_{m+n-3} - e$$

that lacks a blue T_n^*. Denote by G_R and G_B the subgraphs spanned by the red edges and the blue edges, respectively. If $\Delta(G_B) \leq n - 3$, then

$$\delta(G_R) \geq m + n - 5 - (n - 3) = m - 2,$$

and by Lemma 2.1, G_R contains a subgraph isomorphic to T_m^*. So, assume that $\Delta(G_B) \geq n - 2$ and let x be a vertex with blue degree $\deg_{G_B}(x) \geq n - 2$. Pick a vertex set $X \subseteq N_{G_B}(x)$ with cardinality $|X| = n - 2$. Also, define the vertex set

$$Y := V(K_{m+n-3} - e) - (X \cup \{x\}),$$

which has cardinality $|Y| = m - 2$.

Since a blue T_n^* was assumed to be avoided in the red/blue coloring of $K_{m+n-3} - e$ and x joins to X via only blue edges, it follows that all edges joining X and Y must be red. We claim that the subgraph spanned by these red edges must contain a red T_m^*, despite the possibility that one of the edges may be the missing edge. Suppose that the vertices a and b are the endpoints of the missing edge, where $a \in X$ and $b \in Y$. Since $n \geq 5$, X contains some vertex $y \neq a$ which joins to all vertices in Y via red edges. So, the vertex y has red degree at least $m - 2$ Since $m \geq 5$, there exists a vertex $z \neq b$. Noting that az is red, a red T_m^* is formed with vertex y serving as the vertex of degree $m - 2$, vertex z serving as the vertex of degree 2. It follows that $r_*(T_m^*, T_n^*) \leq m + n - 5$. □

2.2 Trees Versus Non-Trees

One of the first evaluations of a star-critical Ramsey number was the case of a tree versus a complete graph. Hook and Isaak (see [47] and [50]) evaluated these numbers after proving that up to isomorphism, the critical colorings for trees versus complete graphs are unique.

2.2.1 Trees Versus Complete Graphs

Letting T_m be a tree of order m, Chvátal proved in 1977 [21] that T_m is n-good:

$$r(T_m, K_n) = (m - 1)(n - 1) + 1.$$

Theorems 1.3, 1.4, and 1.5 all imply that

$$r_*(T_m, K_n) \geq (m - 1)(n - 2) + 1.$$

Hook and Isaak [50] proved that this bound is exact. Their proof depended upon the observation that a critical coloring for (T_m, K_n) has a unique structure. This structure corresponds with the construction that Chvátal and Harary [22] used to obtain the lower bound given in Inequality (1.1).

Theorem 2.9 ([47, 50]) *If* $m, n \geq 2$ *and*

$$c : E(K_{(m-1)(n-1)}) \longrightarrow \{red, blue\}$$

is a critical coloring for (T_m, K_n), *then* c *corresponds with replacing all of the vertices in a blue* K_{n-1} *with copies of a red* K_{m-1}.

Proof Proceed by induction on $n \geq 2$. When $n = 2$, we have that $r(T_m, K_2) = m$ and the critical graph has order $m - 1$. Since it does not contain any blue edges, it must be a red K_{m-1}, which certainly does not contain a red T_m. Assume that the only critical coloring of $(T_m, K_{n'})$ corresponds with replacing all of the vertices in a blue $K_{n'-1}$ with copies of a red K_{m-1} whenever $n' < n$ and let c be a critical coloring for (T_m, K_n), where $n > 2$. Consider two cases.

Case 1 Suppose that the blue degree of every vertex is at most $(m-1)(n-2)-1$. Then the red degree of each vertex is at least

$$((m-1)(n-1)-1) - ((m-1)(n-2)-1) = m-1.$$

By Lemma 1.1, it follows that there exists a red T_m. So, no critical coloring of (T_m, K_n) exists in which every vertex has blue degree at most $(m-1)(n-2)-1$.

Case 2 Suppose that there exists a vertex v with blue degree at least $(m-1)(n-2)$. Let H be the subgraph induced by the blue neighbors of v. By the inductive hypothesis, H contains a red T_m, a blue K_{n-1}, or is a critical coloring for (T_m, K_{n-1}). Since c is assumed to be a critical coloring for (T_m, K_n), H does not contain a red T_m. If H contains a blue K_{n-1}, then including v produces a blue K_n, again contradicting the assumption that c is a critical coloring for (T_m, K_n). So, H must be a critical coloring for (T_m, K_{n-1}), and hence, can be formed by replacing the vertices in a blue K_{n-2} with copies of a red K_{m-1}. If any edge vh, where $h \in V(H)$, is red, then a red T_m is produced. So, all such edges must be blue. Similarly, every edge connecting $K_{(m-1)(n-1)} - H$ to H must also be blue. If any two vertices $u, w \in V(K_{(m-1)(n-1)} - H)$ are joined by a blue edge, then a blue K_n can be formed. It follows that all edges in $K_{(m-1)(n-1)} - H$ must be red, and the critical coloring of (T_m, K_n) is the one described in the statement of Theorem 2.9.

$\square$

The previous theorem can be extended to the case of a tree versus multiple complete graphs. Let $m, n_i \geq 2$ for all $1 \leq i \leq t$. If T_m is any tree of order m, then

$$r(T_m, K_{n_1}, K_{n_2}, \ldots, K_{n_t}) = (m-1)(r(K_{n_1}, K_{n_2}, \ldots, K_{n_t})-1)+1,$$

which follows from Theorem 2.1 of Omidi and Raeisi's paper [73].

Corollary 2.1 *If* $m, n_i \geq 2$ *for all* $1 \leq i \leq t$ *and*

$$c : E(K_{(m-1)(r(K_{n_1}, K_{n_2}, \ldots, K_{n_t})-1)}) \longrightarrow \{1, 2, \ldots, t+1\}$$

is a critical coloring for $(T_m, K_{n_1}, K_{n_2}, \ldots, K_{n_t})$, *then* c *can be formed by replacing all of the vertices in a critical coloring of* $(K_{n_1}, K_{n_2}, \ldots, K_{n_t})$ *in colors* $2, 3, \ldots, t+1$ *with copies of* K_{m-1} *in color* 1.

Proof Let $p = r(K_{n_1}, K_{n_2}, \ldots, K_{n_t})$ and suppose that

$$c : E(K_{(m-1)(r(K_{n_1}, K_{n_2}, \ldots, K_{n_t})-1)}) \longrightarrow \{1, 2, \ldots, t+1\}$$

is a critical coloring for $(T_m, K_{n_1}, K_{n_2}, \ldots, K_{n_t})$. Group together colors $2, 3, \ldots, t+1$ and note that

$$r(T_m, K_p) = r(T_m, K_{n_1}, K_{n_2}, \ldots, K_{n_t}).$$

It follows from Theorem 2.9 that c corresponds with replacing the vertices in a K_{p-1} in colors $2, 3, \ldots, t+1$ with copies of K_{m-1} in color 1. □

The following theorem appeared in [5], generalizing the 2-color result of Hook and Isaak [50].

Theorem 2.10 ([5]) *If* $m, n_i \geq 2$ *for all* $1 \leq i \leq t$ *and* T_m *is a tree of order* m, *then*

$$r_*(T_m, K_{n_1}, K_{n_2}, \ldots, K_{n_t}) = (m-1)(r(K_{n_1}, K_{n_2}, \ldots, K_{n_t}) - 2) + 1.$$

Proof Let $p = r(K_{n_1}, K_{n_2}, \ldots, K_{n_t})$ and note that T_m is p-good. From Theorem 1.3, we find that

$$r_*(T_m, K_{n_1}, K_{n_2}, \ldots, K_{n_t}) \geq r(T_m, K_{n_1}, K_{n_2}, \ldots, K_{n_t}) - r(T_m) + r_*(T_m)$$
$$\geq (m-1)(p-1) + 1 - (m-1)$$
$$\geq (m-1)(p-2) + 1.$$

This inequality also follows from replacing a every vertex in a critical coloring for $(K_{n_1}, K_{n_2}, \ldots, K_{n_t})$ with copies of K_{m-1} in color 1. Introduce a vertex v and select a vertex a in one of the K_{m-1}-subgraphs. Let the edges joining v to the K_{m-1} that contains a be the missing edges. For all other vertices x not contained in the K_{m-1} that includes a, color edge vx the same color as edge ax. The resulting $K_{(m-1)(p-1)} \sqcup K_{(m-1)(p-2)}$ avoids a T_m in color 1 and a K_{n_i} in color $i+1$, for all $1 \leq i \leq t$. For example, see Fig. 2.3.

To prove the reverse inequality, consider a red/blue coloring of $K_{(m-1)(p-1)} \sqcup K_{1,(m-1)(p-2)+1}$ and let v be the center vertex for the missing star. Removing vertex v results in a red/blue coloring of $K_{(m-1)(p-1)}$. If this coloring avoids a T_m in color 1 and a K_{n_i} in color $i+1$, for all $1 \leq i \leq t$, then by Corollary 2.1, it can be formed by replacing all of the vertices in a critical coloring of $(K_{n_1}, K_{n_2}, \ldots, K_{n_t})$ in colors $2, 3, \ldots, t+1$ with copies of K_{m-1} in color 1.

Fig. 2.3 A 3-coloring of $K_{5(m-1)} \sqcup K_{1,4(m-1)}$ that avoids a red T_m, a blue K_3, and a green K_3

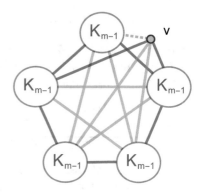

Reintroducing vertex v, if any edge from v to the $K_{(m-1)(p-1)}$ is given color 1, then it joins with a K_{m-1} in color 1 and forms a T_m in color 1, with v being a leaf in T_m. So, all $(m-1)(p-2)+1$ edges joining v to the $K_{(m-1)(p-1)}$ are assigned colors $2, 3, \ldots, t+1$. By the pigeonhole principle, v must join to at least one vertex in each of the $p-1$ copies of K_{m-1}. Let $x_1, x_2, \ldots, x_{p-1}$ be vertices in distinct copies of K_{m-1} such that vx_j is an edge for every j such that $1 \leq j \leq p-1$. Since the subgraph induced by $\{v, x_1, x_2, \ldots, x_{p-1}\}$ is a K_p spanned by edges in colors $2, 3, \ldots, t+1$, there must exist a K_{n_i} in color $i+1$, for some i, where $1 \leq i \leq t$. It follows that

$$r_*(T_m, K_{n_1}, K_{n_2}, \ldots, K_{n_t}) \leq (m-1)(p-2)+1,$$

completing the proof of the theorem.   □

When $t = 1$, Theorem 2.10 states that

$$r(T_m, K_n) = (m-1)(n-1)+1,$$

corresponding with Theorem 2.5 of [50].

Recall that the disjoint union of trees is called a forest. If F is a forest, then let $m(F)$ denote the number of vertices in its largest component. For each $i \in \{1, 2, \ldots, m(F)\}$, let $k_i(F)$ be the number of components in F that have order i. The *variety* of F is defined by

$$C(F) := \{i \mid k_i(F) \neq 0\}$$

and its cardinality is denoted by $q(F) := |C(F)|$. In 1975, Stahl [83] proved that

$$r(F, K_n) = \max_{j \in C(F)} \left\{ (j-1)(n-1) + \sum_{i=j}^{m(F)} ik_i(F) \right\}.$$

Recently, Kamranian and Raeisi [55] extended Stahl's result to the multicolor setting and determined the corresponding star-critical Ramsey number. They proved first that if F is a forest, $n_1, n_2, \ldots, n_t$ are positive integers, and $p = r(K_{n_1}, K_{n_2}, \ldots, K_{n_t})$, then

$$r(F, K_{n_1}, K_{n_2}, \ldots, K_{n_t}) = \max_{j \in C(F)} \left\{ (j-1)(p-2) + \sum_{i=j}^{m(F)} i k_i(F) \right\}.$$

Theorem 2.11 ([55]) *Let F be a forest without isolated vertices, $n_1, n_2, \ldots, n_t$ be positive integers, and $p = r(K_{n_1}, K_{n_2}, \ldots, K_{n_t})$. If j_0 is the smallest value of j for which the maximum value*

$$\max_{j \in C(F)} \left\{ (j-1)(p-2) + \sum_{i=j}^{m(F)} i k_i(F) \right\}$$

is attained, then

$$r_*(F, K_{n_1}, K_{n_2}, \ldots, K_{n_t}) = (j_0 - 1)(p-3) + \sum_{i=j_0}^{m(F)} i k_i(F).$$

2.2.2   Paths Versus Non-Trees

In the case of paths versus graphs other than trees, the following theorem was originally proved by Hook [47]. It is equivalent to Theorem 3.5 of [5] (and corrects a typo in the statement of that theorem).

Theorem 2.12 ([47]) *For all $n \geq 4$, $r_*(P_3, K_n - e) = 2n - 5$.*

Proof It is well known that $r(P_3, K_n - e) = 2n - 3$ (see Section 3.1 of [76]). It follows that P_3 is $(K_n - e)$-good and Theorem 1.3 implies that $r_*(P_3, K_n - e) \geq 2n - 5$. Figure 2.4 also shows a 2-coloring of $K_{2n-4} \sqcup K_{1,2n-6}$ that lacks a red P_3 and a blue $K_n - e$.

In order to prove the inequality $r_*(P_3, K_n - e) \leq 2n - 5$, it must be shown that every red/blue coloring of $K_{2n-3} - e$ contains a red P_3 or a blue $K_n - e$. Consider a red-blue coloring of $K_{2n-3} - e$ and denote by a and b the endpoints of the missing edge. Remove vertex a and assume that the resulting red/blue coloring of K_{2n-4} lacks a red copy of P_3 and a blue copy of $K_n - e$ (otherwise, we are done). Thus, the red edges in this coloring must form a matching M. Let m be the cardinality of M and note that $m \leq n - 2$ since each edge in a matching has two unique vertices. A blue copy of $K_n - e$ can only be formed by taking at most one vertex from $m - 1$ of the edges in M, two vertices from the remaining edge in M, and all vertices that are not incident with a red edge. In order to avoid this, we require

Fig. 2.4 A red/blue coloring of $K_{2n-4} \sqcup K_{1,2n-6}$ that lacks a red P_3 and a blue $K_n - e$

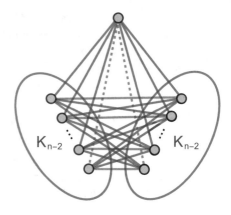

$$1 + m + (2n - 4 - 2m) < n \quad \Longrightarrow \quad m > n - 3.$$

So, a red/blue coloring of K_{2n-4} that lacks a red copy of P_3 and a blue copy of $K_n - e$ must contains a red matching of size $n - 2$. Reintroduce vertex a and note that if a is incident with any red edges, then a red P_3 is formed. If all $2n - 5$ edges incident with a are blue, then a must be adjacent with at least one vertex in each matching, which we label by $x_1, x_2, \ldots, x_{n-2}$. This only accounts for $n - 2$ edges, so there must exist a vertex y such that ay is blue and $x_k y$ is red, for some $1 \leq k \leq n - 2$. Then the subgraph induced by $\{a, y, x_1, x_2, \ldots, x_{n-2}\}$ forms a blue $K_n - e$.   □

The following is Theorem 3.3 of [5]. Note that the graph $K_4 - e$ is the same as the book B_2.

Theorem 2.13 ([5]) $r_*(P_4, K_4 - e) = 4$.

Proof Since $r(P_4, K_4 - e) = 7$ [22], it follows that P_4 is $(K_4 - e)$-good. Thus, Theorem 1.3 implies that $r_*(P_4, K_4 - e) \geq 4$. Figure 2.5 also shows a red/blue coloring of $K_6 \sqcup K_{1,3}$ that lacks a red P_4 and a blue $K_4 - e$, from which this lower bound follows.

In order to prove that $r_*(P_4, K_4 - e) \leq 4$, it must be shown that every red/blue coloring of $K_6 \sqcup K_{1,4}$ contains a red P_4 or a blue $K_4 - e$. Consider a red/blue coloring of $K_6 \sqcup K_{1,4}$ and let a be the center vertex of the missing star. Denote the other vertices in the missing star by b and g. Removing vertices a and g results in a red/blue coloring of K_5. Since $r(P_3, K_4 - e) = 5$ (see Section 3.1 of [76]), there exists is a red P_3 or a blue $K_4 - e$. In the latter case, we are done, so assume there is a red P_3. It is necessary to consider three cases, based on the location of the red P_3 relative to the missing star in the original $K_6 \sqcup K_{1,4}$.

Case 1   Suppose that one of the endpoints of the red P_3 is a leaf in the missing star (see the first image in Fig. 2.6). In this case, avoiding a red P_4 forces edges be, bf, bg, de, df, and dg to be blue. Avoiding a blue $K_4 - e$ then forces bd, ef, eg, and fg to be red (see the second image in Fig. 2.6). If any one of ac, ad, af,

Fig. 2.5 A red/blue coloring of $K_6 \sqcup K_{1,3}$ that lacks a red P_4 and a blue K_3 (and hence, a red $K_{1,3} + e$ and a blue $K_4 - e$)

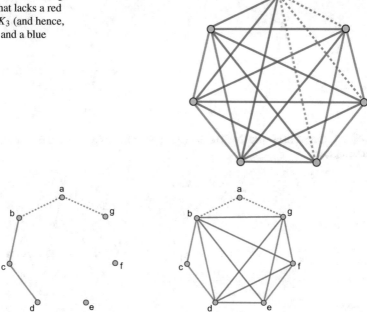

Fig. 2.6 Case 1 in the proof of Theorem 2.13: one of the endpoints in the red P_3 is a leaf in the missing star

or cf are red, then a red P_4 is formed. Otherwise, all four edges are blue and the subgraph induced by $\{a, c, d, f\}$ contains a blue $K_4 - e$.

Case 2 Suppose that the vertex of degree 2 in the red P_3 is a leaf in the missing star (see the first image in Fig. 2.7). In this case, avoiding a red P_4 forces edges ce, cf, de, and df to be blue. Avoiding a blue $K_4 - e$ forces cd to be red (see the second image in Fig. 2.7). At this point, we have reduced this case back to Case 1 (there exists a red P_3 in which one of its endpoints is a leaf in the missing star).

Case 3 Suppose that the red P_3 does not share any vertices with the missing star. (see the first image in Fig. 2.8). Avoiding a red P_4 forces edges $ac, ae, bc, be, cf, cg, ef$, and eg to be blue. Avoiding a blue $K_4 - e$ forces edges af, ce, and fg to be red (see the second image in Fig. 2.8). If bg is red, then a red P_4 is formed. Otherwise, the subgraph induced by $\{b, c, e, g\}$ contains a blue $K_4 - e$.

In all three cases, it has been shown that every red/blue coloring of $K_6 \sqcup K_{1,4}$ contains a red P_4 or a blue $K_4 - e$. It follows that $r_*(P_4, K_4 - e) \leq 4$. ☐

Since $P_3 = K_{1,2}$, a general result by Jacobson [53] concerning stars and a complete graph implies that

$$r(\underbrace{P_3, P_3, \ldots, P_3}_{t \text{ terms}}, K_n) = (r^t(P_3) - 1)(n - 1) + 1,$$

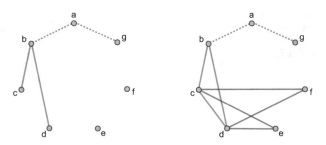

Fig. 2.7 Case 2 in the proof of Theorem 2.13: the vertex of degree 2 in the red P_3 is a leaf in the missing star

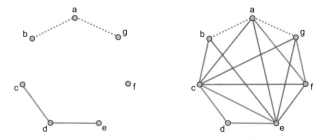

Fig. 2.8 Case 3 in the proof of Theorem 2.13: the red P_3 does not share any vertices with the missing star

for all $n, t \geq 1$. Thus, the multiset consisting of t copies of P_3 is n-good. When $t \geq 1$ is odd, we have $r^t(P_3) = t + 2$ [51] and $r_*^t(P_3) = 1$ (this was proved in Theorem 2.3). Theorem 1.3 then implies that

$$r_*(\underbrace{P_3, P_3, \ldots, P_3}_{t \text{ terms}}, K_n) \geq (r^t(P_3) - 1)(n - 2) + 1.$$

In the $t = 2$ case, $r(P_3, P_3) = 3$ [34] and $r_*(P_3, P_3) = 1$ (this was proved in Theorem 2.2). Theorem 1.3 then implies that

$$r_*(P_3, P_3, K_n) \geq 2n - 2.$$

As the corresponding Ramsey number is only one more than this value, it follows that (P_3, P_3, K_n) is Ramsey-full:

$$r_*(P_3, P_3, K_n) = r(P_3, P_3, K_n) - 1 = 2n - 2.$$

A theorem due to Omidi and Raeisi (see Theorem 2.1 of [73]) then implies that for all $n_i \geq 2$ (where $1 \leq i \leq t$),

$$r(P_3, P_3, K_{n_1}, K_{n_2}, \ldots, K_{n_t}) = (r(P_3, P_3) - 1)(r(K_{n_1}, K_{n_2}, \ldots, K_{n_t}) - 1) + 1$$
$$= 2r(K_{n_1}, K_{n_2}, \ldots, K_{n_t}) - 1.$$

The following theorem from [5] implies that $(P_3, P_3, K_{n_1}, K_{n_2}, \ldots, K_{n_t})$ is Ramsey-full.

Theorem 2.14 ([5]) *For all $1 \le i \le t$ and $n_i \ge 2$,*

$$r_*(P_3, P_3, K_{n_1}, K_{n_2}, \ldots, K_{n_t}) = 2r(K_{n_1}, K_{n_2}, \ldots, K_{n_t}) - 2.$$

Proof The theorem follows from providing a $(t + 2)$-coloring of

$$K_{2r(K_{n_1}, K_{n_2}, \ldots, K_{n_t}) - 1} - e$$

that avoids a monochromatic copy of P_3 in colors 1 and 2 and a copy of K_{n_i} in color $i + 2$, for all $1 \le i \le t$. Let $p = r(K_{n_1}, K_{n_2}, \ldots, K_{n_t})$ and consider t-coloring of K_{p-1} in colors $3, 4, \ldots, t + 2$ that avoids a copy of K_{n_i} in color $i + 2$, for all $1 \le i \le t$. Replace each of the vertices in this K_{p-1} with a copy of K_2 in color 1. Label the vertices in one of the K_2 subgraphs a and b. Introduce a vertex v and color edge vb with color 2. For all vertices $x \ne a, b$, color edge vx the same color as edge ax. The edge joining v to a is the missing edge. The resulting $K_{2p-1} - e$ certainly avoids monochromatic copies of P_3 in colors 1 and 2. No monochromatic complete graph can include both v and x and no monochromatic copy of K_{n_i} in color $i + 2$ exists that include v since no such graph existed that included a. It follows that

$$r_*(P_3, P_3, K_{n_1}, K_{n_2}, \ldots, K_{n_t}) = 2p - 2,$$

from which the theorem follows.   □

Using the currently known classical Ramsey numbers given in Sections 2.1 and 6.1 of Radziszowski's dynamic survey [76], the following star-critical Ramsey numbers follow from Theorem 2.14.

$$r(K_n) = n \implies r_*(P_3, P_3, K_n) = 2n - 2$$
$$r(K_3, K_3) = 6 \implies r_*(P_3, P_3, K_3, K_3) = 10$$
$$r(K_3, K_4) = 9 \implies r_*(P_3, P_3, K_3, K_4) = 16$$
$$r(K_3, K_5) = 14 \implies r_*(P_3, P_3, K_3, K_5) = 26$$
$$r(K_3, K_6) = 18 \implies r_*(P_3, P_3, K_3, K_6) = 34$$
$$r(K_3, K_7) = 23 \implies r_*(P_3, P_3, K_3, K_7) = 44$$
$$r(K_3, K_8) = 28 \implies r_*(P_3, P_3, K_3, K_8) = 54$$

$$r(K_3, K_9) = 36 \quad \Longrightarrow \quad r_*(P_3, P_3, K_3, K_9) = 70$$

$$r(K_4, K_4) = 18 \quad \Longrightarrow \quad r_*(P_3, P_3, K_4, K_4) = 34$$

$$r(K_4, K_5) = 25 \quad \Longrightarrow \quad r_*(P_3, P_3, K_4, K_5) = 48$$

$$r(K_3, K_3, K_3) = 17 \quad \Longrightarrow \quad r_*(P_3, P_3, K_3, K_3, K_3) = 32$$

$$r(K_3, K_3, K_4) = 30 \quad \Longrightarrow \quad r_*(P_3, P_3, K_3, K_3, K_4) = 58$$

2.3 Cycles

This section focuses on star-critical Ramsey numbers involving cycles. We begin first with the case of cycles versus cycles.

2.3.1 Cycles Versus Cycles

The complete evaluation of the cycle-cycle Ramsey number can be found in Faudree and Schelp's paper [32], building off of the prior work of Bondy and Erdős [1] and Chartrand and Schuster [16]. For all $n \geq m \geq 3$,

$$r(C_m, C_n) = \begin{cases} 6 & \text{if } (m, n) = (3, 3) \text{ or if } (m, n) = (4, 4) \\ 2n - 1 & \text{if } m \text{ is odd and } (m, n) \neq (3, 3) \\ n + \frac{m}{2} - 1 & \text{if } m, n \text{ are both even and } (m, n) \neq (4, 4) \\ \max\{n + \frac{m}{2} - 1, 2m - 1\} & \text{if } m \text{ is even and } n \text{ is odd.} \end{cases}$$

In particular, when $m = 3$, Faudree and Schelp [32] proved that for all $n \geq 3$, $r(K_3, C_n) = 2n - 1$. Hook's dissertation [47] included the evaluation of $r_*(C_3, C_n)$ and a classification of the critical colorings for (C_3, C_n). To describe these colorings, first form G_1 by replacing the vertices in a red K_2 with blue copies of K_{n-1}. The coloring G_2 is formed in a similar manner to G_1, except that a single edge joining the two K_{n-1}-subgraphs is colored blue. The resulting colorings of K_{2n-2} are given in Fig. 2.9. Both colorings are in $\mathrm{Crit}(C_3, C_n)$ since the red subgraphs are bipartite and any blue cycle must be entirely contained within one of the blue K_{n-1}-subgraphs (so a blue C_n is not possible). Hook [47] proved that G_1 and G_2 are the only critical colorings of (C_3, C_n).

Theorem 2.15 ([47]) *For all $n \geq 4$, $\mathrm{Crit}(C_3, C_n) = \{G_1, G_2\}$, where G_1 and G_2 are the critical colorings given in Fig. 2.9.*

Proof For $n \geq 4$, consider a red/blue coloring of K_{2n-2} that avoids a red C_3 and a blue C_n. Since $r(C_3, C_{n-1}) = 2(n - 1) - 1 = 2n - 3$ when $n \geq 5$ and $r(C_3, C_3) = 6$ (corresponding to the case $n = 4$), it follows that there

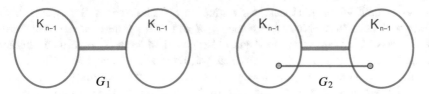

Fig. 2.9 Critical colorings for (C_3, C_n), when $n \geq 4$

exists a blue C_{n-1}, which is assumed to be given by $x_1 x_2 \cdots x_{n-1} x_1$. Denote the other vertices by $y_1, y_2, \ldots, y_{n-1}$. If the subgraph induced by $\{y_1, y_2, \ldots, y_{n-1}\}$ contains a red edge, say $y_1 y_2$ is red. Suppose that y_1 joins to $\{x_1, x_2, \ldots, x_{n-1}\}$ via at least two blue edges, say $y_1 x_i$ and $y_1 x_j$. If $y_1 x_{i+1}$ is also blue (where $x_n = x_1$), then $x_1 x_2 \cdots x_i y x_{i+1} x_{i+2} \cdots x_{n-1} x_1$ is a blue C_n. So, if a blue C_n is avoided, then both $y_1 x_{i+1}$ and $y_1 x_{j+1}$ must be red and i and j differ by more than 1. Without loss of generality, assume that $i + 1 < j$. If $x_{i+1} x_{j+1}$ is blue, then $x_i y_1 x_j x_{j-1} \cdots x_{i+1} x_{j+1} x_{j+2} \cdots x_{n-1} x_1 x_2 \cdots x_i$ is a blue C_n. Otherwise, $x_{i+1} x_{j+1}$ is red and $\{y_1, x_{i+1}, x_{j+1}\}$ forms a red C_3. Therefore, each of y_1 and y_2 is joined with at most one blue edge to $\{x_1, x_2, \ldots, x_{n-1}\}$. Since all of the remaining edges joining $\{y_1, y_2\}$ to $\{x_1, x_2, \ldots, x_{n-1}\}$ are red, by the pigeonhole principle, there exists some i such that $1 \leq i \leq n - 1$ and $y_1 x_i$ and $y_2 x_i$ are both red. Then $\{y_1, y_2, x_i\}$ forms a red C_3.

Based on the previous argument, no red edge exists in $\{y_1, y_2, \ldots, y_{n-1}\}$. Similarly, $y_1 y_2 \cdots y_{n-1} y_1$ is a blue C_{n-1}, so if any red edge exists in $\{x_1, x_2, \ldots, x_{n-1}\}$, then a red C_3 is formed. It follows that the subgraphs induced by $\{x_1, x_2, \ldots, x_{n-1}\}$ and $\{y_1, y_2, \ldots, y_{n-1}\}$ are blue K_{n-1}-subgraphs. Call these two complete subgraphs X an Y, respectively. If any two edges joining X and Y are blue, then a blue C_n can be formed. It follows that at most one blue edge joins X to Y, resulting in the critical colorings described in Fig. 2.9. □

Theorem 2.16 ([47]) *For all $n \geq 3$, $r_*(C_3, C_n) = n + 1$.*

Proof For the lower bound, begin with red K_2 and replace each of its vertices with a blue K_{n-1} (G_1 in Fig. 2.9). The resulting red/blue coloring of K_{2n-2} lacks a red C_3 since the red subgraph is bipartite and it lacks a blue C_n since the largest blue component has order $n - 1$. Add in a vertex v, joining it with all of the vertices in one of the K_{n-1}-subgraphs with red edges and to a single vertex in the other K_{n-1}-subgraph with a blue edge. A red C_3 is still avoided since v is not contained in any red C_3. Since v is only incident with a single blue edge, a blue C_n is still avoided. It follows that $r_*(C_3, C_n) \geq n + 1$.

To prove the reverse inequality, consider a red/blue coloring of $K_{2n-2} \sqcup K_{1,n+1}$ and let v be the center vertex for the missing star. Deleting v results in a red/blue coloring of K_{2n-2}. If a red C_3 and blue C_n are avoided then this coloring must be one of the two colorings described in Theorem 2.15.

For either of these colorings, denote the two blue K_{n-1} subgraphs by X and Y. If v joins to at least two vertices in X via blue edges, then a blue C_n is formed. The same is true if v joins to at least two vertices in Y via blue edges. So, assume that v joins with at most one blue edge to each of X and Y. If v joins to at least one vertex in X and at least one vertex in Y via red edges, then a red C_3 is formed. Since v is incident with at least $n + 1$ edges, it must join to at least two vertices in each of X and Y and at least three vertices in one of X and Y. Thus, v must join to one of X and Y via at least one red edge and to the other set via at least two red edges. Without loss of generality, assume that $x \in X$, $y, z \in Y$, and vx, vy, and vz are all red. Even if one of xy and xz is blue, the other edge is red, forming a red C_3. It follows that $r_*(C_3, C_n) \leq n + 1$. □

In the case where $m = 4$ (see [56]), the cycle-cycle Ramsey number simplifies to

$$r(C_4, C_n) = \begin{cases} 7 & \text{if } n = 3, 5 \\ 6 & \text{if } n = 4 \\ n + 1 & \text{if } n \geq 6. \end{cases}$$

Theorem 2.16 implies that $r_*(C_4, C_3) = {}^{\cdot}5$. In 2015, Wu et al. [90] extended this result to all cases of C_4 versus C_n, when $n \geq 4$. The case of $n = 4$ is given in Theorem 2.17 below. It implies that (C_4, C_4) is Ramsey-full.

Theorem 2.17 ([90]) $r_*(C_4, C_4) = 5$.

Proof Chartrand and Schuster [16] proved that $r(C_4, C_4) = 6$. To prove this theorem, a red/blue coloring of $K_6 - e$ that avoids a monochromatic C_4 must be constructed. Begin with a red/blue K_5 in which the subgraphs spanned by each color are isomorphic to C_5. Let $x_1x_2x_3x_4x_5x_1$ be the red cycle and $x_1x_3x_5x_2x_4x_1$ be the blue cycle. Introduce vertex v, joining it to vertices x_3 and x_4 with red edges and vertices x_2 and x_5 with blue edges. The edge joining x_1 to v is the missing edge (see Fig. 2.10). The resulting $K_6 - e$ lacks a monochromatic C_4, from which it follows that $r_*(C_4, C_4) \geq 5$. Since $r_*(C_4, C_4) \leq r(C_4, C_4) - 1$, it follows that $r_*(C_4, C_4) = 5$. □

In the cases $n = 5, 6$, Chartrand and Schuster [16] proved that

Fig. 2.10 A red/blue coloring of $K_6 - e$ that lacks a monochromatic C_4

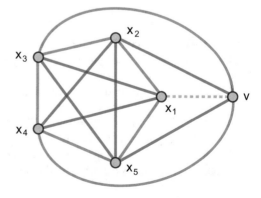

Fig. 2.11 A red/blue
coloring of $K_6 \sqcup K_{1,4}$ that
lacks a red C_4 and a blue C_5

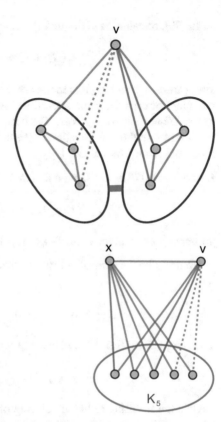

Fig. 2.12 A red/blue
coloring of $K_6 \sqcup K_{1,4}$ that
lacks a red C_4 and a blue C_6

$$r(C_4, C_5) = r(C_4, C_6) = 7.$$

When $n = 5$, begin by replacing the vertices in a blue K_2 with copies of red K_3-subgraphs. Introduce vertex v, joining it to one of the K_3-subgraphs using two blue edges and one red edge. Also, join v to the other K_3-subgraph using a single red edge (the two other edges are the missing edges). The resulting $K_6 \sqcup K_{1,4}$ is shown in Fig. 2.11. It avoids a red C_4 and all of its blue cycles have even length, so a blue C_5 is also avoided. It follows that $r_*(C_4, C_5) \geq 5$.

When $n = 6$, consider a blue K_5, with vertex set $\{y_1, y_2, y_3, y_4, y_5\}$, and introduce a vertex x, joining it to y_1 with a blue edge and to y_2, y_3, y_4, and y_5 with red edges. Next, introduce vertex v, joining it to y_1 and y_2 with red edges, joining it to y_3 with a blue edge, and letting the edges joining it to y_4 and y_5 be the missing edges. See Fig. 2.12. The resulting $K_6 \sqcup K_{1,4}$ avoids a red C_4 and a blue C_6, from which it follows that $r_*(C_4, C_6) \geq 5$.

In [90], the authors claim that

$$r_*(C_4, C_5) = r_*(C_4, C_6) = 5,$$

after carefully checking all small cases. As they omitted the details of the proofs of the upper bounds for these star-critical Ramsey numbers, the reader is encouraged to formulate their own proofs. The case where $n \geq 7$ was proved in [90], and the proof relies on the classification of the critical colorings for (C_4, C_n), as well as the following lemma due to Ore [74].

Lemma 2.3 ([74]) *Let G be a 2-connected graph of order n. If*

$$\deg_G(x) + \deg_G(y) \geq n + 1$$

for every pair of distinct vertices $x, y \in V(G)$, then G is Hamiltonian-connected.

To describe the critical colorings when $n \geq 7$, begin with a vertex x and form a red

$$G_R := \{x\} + ((n - 2i - 2)K_1 \cup iK_2).$$

The complement forms the blue subgraph

$$G_B := \{x\} \cup (K_{n-2i-2} + (K_{2i} - iK_2)).$$

Denote the resulting red/blue coloring of K_{n-1} by G. For $0 \leq i \leq \frac{n-2}{2}$, form G_1 by adding in a vertex y and joining it to all vertices in G using blue edges. For $0 \leq i \leq \frac{n-2}{2}$, form G_2 by adding in a vertex y, joining it by a red edge to a single vertex in the red $(n-2i-2)K_1$ and by blue edges to the remaining vertices in G. For $1 \leq i \leq \frac{n-2}{2}$, form G_3 by adding in a vertex y, joining it by a red edge to a single vertex in the red iK_2 and by blue edges to the remaining vertices in G. For the last coloring, let $0 \leq i \leq \frac{n-1}{2}$ and form G_4 with a red graph $\{x\} + ((n - 2i - 1)K_1 \cup iK_2)$ and blue graph $\{x\} \cup (K_{n-2i-1} + (K_{2i} - iK_2))$. Figure 2.13 illustrates each of these colorings.

None of the colorings in Fig. 2.13 contain a red C_4 as such a cycle would have to include vertex x and the only cycles that include x have length 3. Also, no blue C_n exists as such a cycle would have to include all n of the vertices and x is an isolated vertex in each of the subgraphs spanned by the blue edges. So, all of these colorings are in Crit(C_4, C_n). The fact that these are the only colorings in Crit(C_4, C_n) was proved by Wu et al. [90].

Theorem 2.18 ([90]) *For all $n \geq 6$, every critical coloring of (C_4, C_n) has the form of one of the colorings G_1, G_2, G_3, G_4 described above (and shown in Fig. 2.13).*

Theorem 2.19 ([90]) *For all $n \geq 7$, $r_*(C_4, C_n) = 5$.*

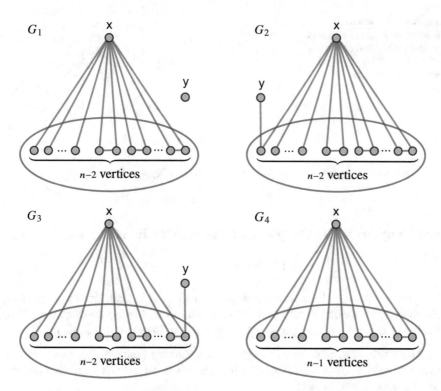

Fig. 2.13 Critical colorings for (C_4, C_n), when $n \geq 6$

Proof Assume that $n \geq 7$ and consider the red/blue coloring of K_n in which the subgraph spanned by red edges is isomorphic to $K_{1,n-2} \cup \{y\}$ and the subgraph spanned by blue edges is isomorphic to $\overline{K_{1,n-2}} + \{y\}$. This coloring corresponds with G_1 in Fig. 2.13 when $i = 0$. Denote by x the center vertex in the red $K_{1,n-2}$ and let $x_1, x_2, \ldots, x_{n-2}$ denote its leaves. Let y be the isolated vertex in the red subgraph. Introduce a new vertex v, joining it to x, y, and x_1 via red edges and to x_2 via a blue edge. The resulting $K_n \sqcup K_{1,4}$ avoids a red C_4 and a blue C_n, from which it follows that $r_*(C_4, C_n) \geq 5$.

Now consider a red/blue coloring of $K_n \sqcup K_{1,5}$ and let v be the center vertex of the deleted star. Removing v results in a red/blue coloring of K_n that, if a red C_4 and a blue C_n are to be avoided, must be one of the colorings described in Fig. 2.13 by Theorem 2.18. Denote by G' this red/blue coloring of K_n. In all four cases, note that the subgraph of G' spanned by red edges contains a red $K_{1,n-2}$, with center vertex given by x. Denote by $x_1, x_2, \ldots, x_{n-2}$ the leaves of the red $K_{1,n-2}$. If a red C_4 is to be avoided, then v joins to $\{x_1, x_2, \ldots, x_{n-2}\}$ via at most one red edge.

As v is assumed to be incident with five edges, it must join to at least three vertices in $\{x_1, x_2, \ldots, x_{n-2}\}$. At least two such edges must be blue, say vx_1 and vx_2. Note that in the blue subgraph induced by $\{x_1, x_2, \ldots, x_{n-2}\}$, which we denote

Fig. 2.14 A red/blue
coloring of $K_{2n-2} \sqcup K_{1,n}$ that
lacks a red C_m and a blue C_n,
where m is odd, $n \geq m \geq 3$,
and $(m, n) \neq (3, 3)$

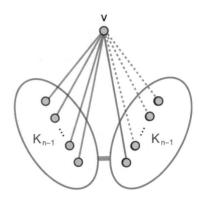

by B, every vertex has blue degree at least $n - 4$. Hence, for any $1 \leq i < j \leq n - 2$,

$$\deg_B(x_i) + \deg_B(x_j) \geq 2n - 8 \geq n - 1$$

is equivalent to $n \geq 7$. Lemma 2.3 then implies that B is Hamiltonian-connected.
So, there exists a Hamiltonian path from x_1 to x_2 in B. Joining this blue path with
the blue edges vx_1 and vx_2 produces a blue C_n. It follows that $r_*(C_4, C_n) \leq 5$.   □

At present, few other cases of star-critical Ramsey numbers for cycles versus
cycles have been considered. One exception to this is the following theorem proved
by Zhang et al. [95] in 2016.

Theorem 2.20 ([95]) *For m odd, $n \geq m \geq 3$, and $(m, n) \neq (3, 3)$,*

$$r_*(C_m, C_n) = n + 1.$$

Observe that under the assumptions of Theorem 2.20, $r(C_m, C_n) = 2n - 1$ [32].
The lower bound for the corresponding star-critical Ramsey number comes from
first replacing the vertices in a red K_2 with blue K_{n-1}-subgraphs. Introduce vertex
v, joining it to all of the vertices in one K_{n-1}-subgraph with red edges and joining it
to a single vertex in the other K_{n-1}-subgraph with a blue edge. The remaining $n - 2$
edges are the missing edges (see Fig. 2.14). The resulting $K_{2n-2} \sqcup K_{1,n}$ lacks a blue
C_n and all of its red cycles have even length. It follows that $r_*(C_m, C_n) \geq n + 1$.

2.3.2   Cycles Versus Other Graphs

In 1989, Burr et al. [12] proved that for all $n \geq 3$,

$$r(C_4, P_n) = n + 1.$$

Fig. 2.15 A critical coloring
for (C_4, P_n) in which a red
matching of size 2 is removed
from the blue K_{n-1}

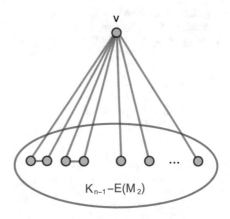

The corresponding star-critical Ramsey number was considered by Hook and Isaak
(see [47] and [50]). Before evaluating $r_*(C_4, P_n)$, it is necessary to describe the
critical colorings of (C_4, P_n).

For $i \in \{0, 1, 2, \ldots, \lfloor \frac{n-1}{2} \rfloor\}$, let M_i denote a set of i independent red edges
($i = 0$ corresponds with the empty set). Let X_i be the graph formed by taking a
K_{n-1} and removing edges from it that correspond to an M_i-subgraph. Now let $G_{i,n}$
be the red/blue coloring of K_n whose red subgraph is $\{v\} + (iK_2 \cup (n-1-2i)K_1)$
and whose blue subgraph is $X_i \cup \{v\}$. For example, Fig. 2.15 shows $G_{2,n}$.

Theorem 2.21 ([47, 50]) *For $n \geq 3$,*

$$\mathrm{Crit}(C_4, P_n) = \left\{ G_{i,n} \; \middle| \; i \in \left\{0, 1, \ldots, \left\lfloor \frac{n-1}{2} \right\rfloor\right\} \right\}.$$

Proof When $n = 3$, the only critical colorings for (C_4, P_3) are the red/blue
colorings of K_3 that have at most one blue edge. The two resulting colorings
correspond with $G_{0,3}$ (having one blue edge) and $G_{1,3}$ (having no blue edges).

For the cases where $n \geq 4$, consider a red/blue coloring of K_n that lacks a red
C_4 and a blue P_n. Since $r(C_4, P_{n-1}) = n$ [12], it follows that there exists a blue
P_{n-1}, which we assume has vertex sequence $x_1 x_2 \cdots x_{n-1}$. Let v be the vertex not
included in the blue path. Since a blue P_n is avoided, the edges vx_1 and vx_{n-1}
must be red. If vx_i and vx_{i+1} are both blue for some $i \in \{2, 3, \ldots, n-3\}$, then
$x_1 x_2 \cdots x_i v x_{i+1} x_{i+2} \cdots x_{n-1}$ is a blue P_n. So, v cannot join to two consecutive
vertices in $x_1 x_2 \cdots x_{n-1}$ via blue edges. Suppose that vx_k is blue for some $k \in$
$\{2, 3, \ldots, n-2\}$. Then vx_{k-1} and vx_{k+1} are both red. Then $x_1 x_{n-1}$ must be
red, otherwise $vx_k x_{k+1} \cdots x_{n-1} x_1 x_2 \cdots x_{k-1}$ is a blue P_n. If $x_{k-1} x_{n-1}$ is red, then
$vx_1 x_{n-1} x_{k-1} v$ is a red C_4. If $x_{k-1} x_{n-1}$ is blue, then $x_1 x_2 \cdots x_{k-1} x_{n-1} x_{n-2} \cdots x_k v$
is a blue P_n. The previous two statements hold for $k = 2$ by symmetry with
$k = n-2$. It follows that v joins to all of the vertices in the blue P_n via red edges.

Suppose that some x_k has red degree $\deg_R(x_k) \geq 3$. That is, vx_k is red and
there exists two distinct vertices x_i and x_j such that $x_i x_k$ and $x_j x_k$ are red. Then

Fig. 2.16 A red/blue
coloring of $K_n \sqcup K_{1,2}$ that
avoids a red C_4 and a blue P_n

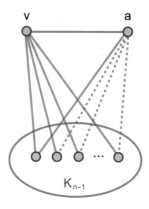

$v x_i x_k x_j v$ is a red C_4. So, each x_k has at most one red edge to another vertex in the P_{n-1}. Hence, the red subgraph induced by $\{x_1, x_2, \ldots, x_{n-1}\}$ is isomorphic to $i K_2 \cup (n - 1 - 2i) K_1$ for some $i \in \{0, 1, \ldots, \lfloor \frac{n-1}{2} \rfloor\}$, and the coloring corresponds with $G_{i,n}$.   □

Theorem 2.22 ([47, 50]) *For all $n \geq 3$, $r_*(C_4, P_n) = 3$.*

Proof To prove that 3 is a lower bound for $r_*(C_4, P_n)$, start with a red K_2 and replace one of its vertices with a blue K_{n-1}. Select a vertex x in the blue K_{n-1} and label the vertex not replaced by a blue K_{n-1} by v. The resulting red/blue K_n is a critical coloring for (C_4, P_n) since no red C_4 exists and the largest connected blue component has order $n - 1$. Now, add in a vertex a and the edges ax_1 and av. Coloring ax_1 red and av blue avoids a red C_4 and a blue P_n (see Fig. 2.16). Thus, $r_*(C_4, P_n) \geq 3$.

To prove the reverse inequality, consider a red/blue coloring of $K_n \sqcup K_{1,3}$ and let w be the center vertex for the missing star. Removing w results in a red/blue coloring of K_n that, if it avoids a red C_4 and a blue P_n, must be one of the colorings described in Theorem 2.21. A blue edge from w to any vertex other than v in the K_n produces a blue P_n and any two edges from v to vertices other than v in the K_n produce a red C_4. Therefore, three edges (including edge vw) guarantees the existence of a red C_4 or a blue P_n. Hence, $r_*(C_4, P_n) \leq 3$.   □

In 1973, Lawrence [58] showed that for all $m, n \in \mathbb{N}$ that satisfy $m \geq 2n$, $r(C_m, K_{1,n}) = m$. The corresponding star-critical Ramsey number was considered in [88].

Theorem 2.23 ([88]) *If $m, n \in \mathbb{N}$ such that $m \geq 2n+2$, then $r_*(C_m, K_{1,n}) = n+1$.*

Proof To prove the lower bound, begin with a red K_{m-1}, add in a vertex v, and join v to K_{m-1} using only n edges. Color of the n edges red and the others blue obtain a red/blue coloring of $K_{m-1} \sqcup K_{1,n}$. No red cycle can include vertex v since it has red degree 1 and there are not enough vertices in the red K_{m-1} to contain a red C_m. Also, no blue $K_{1,n}$ exists since there are only $n - 1$ blue edges. It follows that $r_*(C_m, K_{1,n}) \geq n + 1$.

To prove the reverse inequality, consider a red/blue coloring of $K_{m-1} \sqcup K_{1,n+1}$ and denote by v the center vertex of the missing star. If the coloring lacks a blue $K_{1,n}$, then every vertex has blue degree at most $n - 1$. So, the vertex v must have red degree at least 2. If u is any vertex other than v, then its red degree satisfies

$$\deg_R(u) \geq (m - 2) - (n - 1) = m - n - 1 \geq \left\lfloor \frac{n+1}{2} \right\rfloor.$$

Thus, there exists a degree sequence $d_1 \leq d_2 \leq \cdots \leq d_n$, where $d_1 \geq 2$ and $d_i \geq \lfloor \frac{n+1}{2} \rfloor \geq i + 1$, for $1 < i < \frac{n}{2}$. By Lemma 2.2, there exists a red C_m. It follows that $r_*(C_m, K_{1,n}) \leq n + 1$. $\qquad\square$

The study of Ramsey numbers for cycles versus complete graphs was initiated in the work of Bondy and Erdős [1] in 1973. Three years later, Erdős et al. [28] conjectured that for all $m \geq n$, $r(C_m, K_n) = (m - 1)(n - 1) + 1$ (in other words, C_m is conjectured to be n-good). Their conjecture is still open, although it has been proven for numerous complete graphs and for sufficiently large cycles. In particular, when $n = 3$, Faudree and Schelp [32] proved that for all $m \geq 3$, $r(C_m K_3) = 2m-1$. Since $K_3 = C_3$, Theorem 2.16 implies that $r_*(C_m, K_3) = m + 1$, for all $m \geq 3$.

It was shown by Yang et al. [93] that $r(C_m, K_4) = 3m - 2$ for all $m \geq 4$. In the case $m = 4$, this reduces to $r(C_4, K_4) = 10$. The following theorem implies that (C_4, K_4) is Ramsey-full.

Theorem 2.24 ([54]) $r_*(C_4, K_4) = 9$.

Proof This theorem will follow from providing a red/blue coloring of $K_{10} - e$ that avoids a red C_4 and a blue K_4. Begin with the red graph G given in the first image in Fig. 2.17. One can confirm that G does not contain C_4 as a subgraph. Likewise, $\overline{G}$ does not contain K_4 as a subgraph. Together, G and $\overline{G}$ form a red/blue coloring of K_9 that avoids a red C_4 and a blue K_4. Introduce a vertex v and join v to eight of the vertices in G using the colors red and blue as demonstrated in the second image

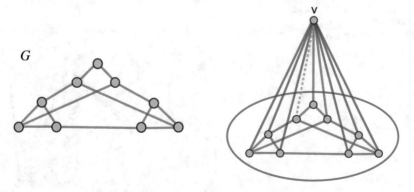

Fig. 2.17 A red/blue coloring of $K_{10} - e$ that avoids a red C_4 and a blue K_4

in Fig. 2.17. It follows that $r_*(C_4, K_4) \geq 9$. Since $r_*(C_4, K_4) \leq r(C_4, K_4) - 1 = 9$, the statement of the theorem follows.   □

Using the following theorem (which follows from Lemmas 1 and 4 in [54]) describing the structure of critical colorings for C_m, K_4), one can determine the value of the star-critical Ramsey number $r_*(C_m, K_4)$ when $m \geq 5$. Although originally considered by Hook [47], the approach used by Jayawardene et al. [54] is described here.

Theorem 2.25 ([47, 54]) *For every* $m \geq 5$, *every red/blue coloring of* K_{3m-3} *in* Crit(C_m, K_4) *contains a red* $3K_{m-1}$.

Theorem 2.26 ([47, 54]) *For all* $m \geq 5$, $r_*(C_m, K_4) = 2m$.

Proof Begin by replacing the vertices in a blue K_3 with copies of red K_{m-1}-subgraphs. The resulting graph does not contain a red C_m since the largest red component has order $m - 1$. It also does not contain a blue K_4 since every blue complete subgraph contains at most a single vertex from any one of the red K_{m-1}-subgraphs. Introduce vertex v and join v to all vertices in two of the red K_{m-1}-subgraphs with blue edges. Join v to a single vertex in the third red K_{m-1}-subgraph with a red edge. The remaining $m - 2$ edges are the missing edges (see Fig. 2.18). The vertex v is not contained in any red cycle since it has red degree 1 and v is not contained in any blue K_4 since such a graph would have to contain v and a single vertex from each of the red K_{m-1}-subgraphs. It follows that $r_*(C_m, K_4) \geq 2m$.

Now consider a red/blue coloring of $K_{3n-3} \sqcup K_{1,2m}$ and let v be the center vertex for the missing star. Removing v results in a red/blue coloring of K_{3m-3}, and if a red C_m and blue K_4 are avoided, then it must contain a red $3K_{m-1}$ by Theorem 2.26. Let X, Y, and Z denote the vertex sets of the three red K_{m-1}-subgraphs. If v joins to any two vertices in one of the red K_{m-1}-subgraphs, say X, via red edges, then a red C_m is formed. So, v can join to at most one vertex in each of X, Y, and Z with

Fig. 2.18 A red/blue coloring of $K_{3m-3} \sqcup K_{1,2m-1}$ that avoids a red C_m and a blue K_4, where $m \geq 5$

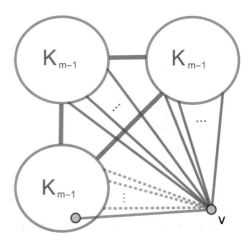

a red edge. Also, note that at most one red edge can exist between any two of the vertex sets X, Y, and Z, otherwise a red C_m is formed.

Case 1   Suppose that v is adjacent to exactly two vertices in one of the sets X, Y, or Z. Then v must be adjacent to all of the vertices in the other two vertex sets. Without loss of generality, assume that v is adjacent to all of the vertices in $X \cup Y$. In particular, v must join to at least four vertices in each of X and Y. Select $x \in X$ and $y \in Y$ such that x does not join via a red edge to any vertex in $Y \cup Z \cup \{v\}$ and y does not join via a red edge to any vertex in $X \cup Z \cup \{v\}$. This is possible since at most three vertices within a given red K_{m-1} can join to vertices outside that K_{m-1} via a red edge. Since no red C_m exists, there exists a vertex $z \in Z$ such that vz is blue and $\{v, x, y, z\}$ forms a blue K_4.

Case 2   Suppose that v is adjacent to at least three vertices in each of X, Y, and Z. Without loss of generality, assume that v is adjacent to at least four vertices in X, at least three vertices in Y, and at least three vertices in Z. Since v can join to at most one vertex in each of Y and Z with a red edge, it is possible to select two vertices $y \in Y$ and $z \in Z$ such that $\{v, y, z\}$ forms a blue K_3. Next, select $x \in X$ that only joins to vertices in $Y \cup Z \cup \{v\}$ with blue edges. This is possible since X can have at most three vertices that join to vertices outside of X with red edges. Then $\{v, x, y, z\}$ forms a blue K_4.

In both cases, every red/blue coloring of $K_{3n-3} \sqcup K_{1,2m}$ contains either a red C_m or a blue K_4. It follows that $r_*(C_m, K_4) \leq 2m$.   □

2.4 Wheels

In 1983, Burr and Erdős [11] proved that for all $n \geq 5$,

$$r(K_3, W_n) = 2n - 1.$$

In her dissertation, Hook [47] considered the star-critical Ramsey number $r_*(K_3, W_n)$. Before presenting Hook's proof, we define a class of graphs that will serve as the critical colorings for (K_3, W_n). A full classification of the critical colorings for (K_3, W_n) was determined by Radziszowski and Jin in [77]. Although this full classification is not necessary to determine $r_*(K_3, W_n)$, the following theorem will be fundamental.

Theorem 2.27 ([77]) *For all $n \geq 6$, if a red/blue coloring of K_{2n-2} is a critical coloring for (K_3, W_n), then it has a blue subgraph isomorphic to K_{n-1}.*

The following lemma will also be useful.

Lemma 2.4 ([47]) *For all $n \geq 4$, the graph $K_{n-1} \sqcup K_{1,3}$ contains a subgraph that is isomorphic to W_n.*

Proof When $n = 4$, it is clear that $K_3 \sqcup K_{1,3} = K_4 = W_4$. For $n \geq 5$, consider the graph $K_{n-1} \sqcup K_{1,3}$ with vertex set $\{v, x_1, x_2, \ldots, x_{n-1}\}$, where v is the vertex of degree 3. Without loss of generality, assume that v is adjacent to x_1, x_2, and x_3. Then $x_2 v x_3 \cdots x_{n-1} \cdots x_2$ forms a C_{n-1} and x_1 is adjacent to every vertex in this cycle, forming a wheel W_n.   □

When $n = 4$, $W_4 = K_4$ and $r_*(K_3, K_4) = 8$ follows from Theorem 1.1 and the fact that $r(K_3, K_4) = 9$ (see [37]). In the following theorem, the star-critical Ramsey number $r_*(K_3, W_n)$ is considered for $n \geq 7$.

Theorem 2.28 ([47]) *For all $n \geq 7$, $r_*(K_3, W_n) = n + 2$.*

Proof Consider the red/blue coloring of K_{2n-2} with red subgraph $K_{n-1,n-1}$ and blue subgraph $2K_{n-1}$. Add in a vertex v, joining it to all of the vertices in one of the blue K_{n-1}-subgraphs using red edges. Join v to exactly two vertices in the other blue K_{n-1} using blue edges (see Fig. 2.19). The red subgraph of the resulting red/blue $K_{2n-2} \sqcup K_{1,n+1}$ is bipartite, and hence, avoids a red K_3. No blue W_n exists since v does not have a large enough blue degree to be contained in a wheel and not enough vertices exist in the blue K_{n-1}-subgraphs to contain W_n. It follows that $r_*(K_3, W_n) \geq n + 2$. Note that this inequality also follows from Lemma 1.3 after observing that W_n is 3-good.

To prove the reverse inequality, consider a red/blue coloring of $K_{2n-2} \sqcup K_{1,n+2}$ and let v denote the center vertex of the missing star. Removing v, we obtain a red/blue coloring of K_{2n-2}. If this coloring avoids a red K_3 and a blue W_n, then by Theorem 2.27, it must contain a blue subgraph isomorphic to K_{n-1} that we denote by X. Let Y denote the subgraph induced by the vertices not in $X \cup \{v\}$ (i.e., Y is a red/blue K_{n-1}) and suppose that $y_1, y_2 \in Y$ are such that $y_1 y_2$ is red. Let y_1 and y_2 have blue degrees d_1 and d_2 in X, respectively. By Lemma 2.4, if either $d_1 \geq 3$ or $d_2 \geq 3$, there exists a blue W_n. So, assume $d_1 \leq 2$ and $d_2 \leq 2$. Then y_1 and y_2 each have red degree at least $n - 3$ in X. Since $2(n - 3) > n - 1$ for all $n \geq 6$, there exists some vertex $x \in X$ such that the subgraph induced by $\{x, y_1, y_2\}$ is a red K_3, which is a contradiction. So, no red edge exists in Y and the critical coloring contains a blue $2K_{n-1}$.

Fig. 2.19 A red/blue coloring of $K_{2n-2} \sqcup K_{1,n+1}$ that avoids a red K_3 and a blue W_n

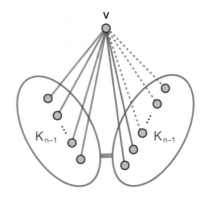

Observe that v joins to at least three vertices in each of X and Y. By Lemma 2.4, if v joins to either of X or Y via three or more blue edges, then a blue W_n is formed. So, assume that v joins to each of X and Y using at most two blue edges (and at least one red edge). Since $n \geq 7$, it follows that v is incident with at least five vertices in one of X or Y, say Y. Then v joins to at least three vertices in Y via red edges. Without loss of generality, assume that vx, vy_1, vy_2, and vy_3 are red, where $x \in X$ and $y_1, y_2, y_3 \in Y$. If xy_1, xy_2, and xy_3 are all blue, then Lemma 2.4 implies that there exists a blue W_n. Otherwise, some xy_i is red, say xy_1, and the subgraph induced by $\{v, x, y_1\}$ forms a red K_3. It follows that $r_*(K_3, W_n) \leq n + 2$. $\square$

Noting that $W_4 = K_4$, it was proved by Yang et al. [93] that

$$r(C_m, W_4) = 3m - 2,$$

when $m \geq 4$. In Theorem 2.26, a result equivalent to

$$r_*(C_m, W_4) = 2m,$$

was proved for all $m \geq 5$. For other wheels, it was proved by Chen et al. [17] that

$$r(C_m, W_n) = 3m - 2,$$

for n even, $m + 1 \geq n \geq 4$, and $m \geq 4$. Before considering the corresponding star-critical Ramsey number, we state the following theorem by Liu and Chen [67], which introduces the requirement that $m \geq 60$.

Theorem 2.29 ([67]) *For n even, $m + 1 \geq n \geq 6$, and $m \geq 60$, every critical coloring of (C_m, W_n) contains a red $3K_{m-1}$.*

In 2021, Liu and Chen [67] proved the following theorem, evaluating the star-critical Ramsey numbers for the cycles and wheels described in Lemma 2.29.

Theorem 2.30 ([67]) *For n even, $m + 1 \geq n \geq 6$, and $m \geq 60$, $r_*(C_m, W_n) = 2m$.*

Proof Begin by replacing the vertices in a blue K_3 with red K_{m-1}-subgraphs to obtain a red/blue coloring of K_{3m-3}. Introduce a vertex v and and join it to all vertices in two of the K_{m-1} subgraphs via blue edges. Join v to a single vertex in the third K_{m-1}-subgraph via a red edge. The remaining $m - 2$ edges are the missing edges. The resulting $K_{3m-3} \sqcup K_{1,2m-1}$ lacks a red C_m since v is not contained in any red cycles and the largest red component in the original K_{3m-3} only had order $m - 1$. The subgraph panned by the blue edges can be properly vertex colored using only three colors since one color can be assigned to each of the K_{m-1}-subgraphs and v can receive the same color as the vertex v joins via a red edge. For n even, $\chi(W_n) = 4$, so no blue W_n-subgraph exists. It follows that $r_*(C_m, W_n) \geq 2m$.

To prove the reverse inequality, consider a red/blue coloring of $K_{3m-3} \sqcup K_{1,2m}$ and let G_R and G_B be the subgraphs spanned by the red and blue edges, respectively. Let v be the center vertex for the missing star. Then deleting v results in a red/blue coloring of K_{3m-3}. If a red C_m and a blue W_n are avoided, then Theorem 2.29 implies that it must contain a red $3K_{m-1}$. Let X_1, X_2, X_3 denote the vertex sets for the three red K_{m-1} subgraphs. For each i such that $1 \le i \le 3$, denote by $N_{G_R}(v, X_i)$ the subset of X_i joined by red edges to v and $\deg_{G_R}(v, X_i) := |N_{G_R}(v, X_i)|$. Define $N_{G_B}(v, X_i)$ and $\deg_{G_B}(v, X_i)$ analogously. If no red C_m exists, then there is at most one red edge between any pair of sets X_1, X_2, X_3 and $\deg_{G_R}(v, X_i) \le 1$ for all $1 \le i \le 3$. It follows that $\deg_{G_B}(v, X_i) \ge 1$ for all $1 \le i \le 3$ and

$$\deg_{G_B}(v, X_1) + \deg_{G_B}(v, X_2) + \deg_{G_B}(v, X_3) \ge 2m - 3.$$

Consider two cases.

Case 1 If $\deg_{G_B}(v, X_3) = 1$, then $\deg_{G_B}(v, X_1) \ge m - 2$ and $\deg_{G_B}(v, X_2) \ge m - 2$. Let $z \in N_{G_B}(v, X_3)$,

$$Y_1 := N_{G_B}(v, X_1) \cap N_{G_B}(z, X_1), \quad \text{and} \quad Y_2 := N_{G_B}(v, X_2) \cap N_{G_B}(z, X_2).$$

Then $|Y_1|, |Y_2| \ge m - 3 \ge \frac{n-2}{2}$. Since the subgraph of G_B induced by $Y_1 \cup Y_2$ is a complete bipartite graph with at most one edge missing, it follows that $Y_1 \cup Y_2 \cup \{v\}$ contains a C_{n-1}, from which a W_n can be formed by including vertex z.

Case 2 If $\deg_{G_B}(v, X_3) \ge 2$, then $\deg_{G_B}(v, X_1) \ge \frac{2m-3}{3}$ and $\deg_{G_B}(v, X_2) \ge \frac{n-2}{2}$. Let $x, y \in N_{G_B}(v, X_3)$. At most one red edge exists between $\{x, y\}$ and $N_{G_B}(v, X_2)$, so assume that $N_{G_B}(v, X_2) \subseteq N_{G_B}(x, X_2)$. Let

$$Z_1 := N_{G_B}(v, X_1) \cap N_{G_B}(x, X_1) \quad \text{and} \quad Z_2 := N_{G_B}(v, X_2).$$

Then $|Z_1| \ge \frac{2n-3}{3} - 1 \ge \frac{n-2}{2}$ and $|Z_2| \ge \frac{m-1}{2} \ge \frac{n-2}{2}$. The subgraph of G_B induced by $Z_1 \cup Z_2$ is a complete bipartite graph with at most one edge missing. It follows that $Z_1 \cup Z_2 \cup \{v\}$ contains a C_{n-1}, from which a W_n can be formed by including vertex x.

In both cases, it has been shown that a red/blue coloring of $K_{3m-3} \sqcup K_{1,2m}$ contains a red C_m or a blue W_n. It follows that $r_*(C_m, W_n) \le 2m$.   □

Recently, Liu and Chen [68] also proved a result equivalent to the following theorem.

Theorem 2.31 ([68]) *For odd $m \ge 5$ and $n \ge \frac{3(m-1)}{2} + 3$, $r_*(C_m, W_n) = n + 2$.*

2.5 Disjoint Unions of Complete Graphs

The first star-critical Ramsey numbers involving disjoint unions of complete graphs were investigated by Hook and Isaak (see [47] and [50]), and concerned disjoint unions of K_2 and K_3. We start with the the case of of matchings mK_2, which are

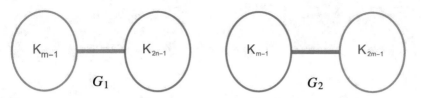

Fig. 2.20 For all $n \geq m \geq 1$, G_1 is a red/blue coloring of K_{m+2n-2} that avoids a red mK_2 and a blue nK_2. When $n = m$, G_2 provides a second red/blue coloring of K_{m+2n-2} that avoids a monochromatic mK_2

sometimes referred to as "stripes" in the literature (e.g., see [23, 24], and [69]). For all $n \geq m \geq 1$, Cockayne and Lorimer [24] proved that

$$r(mK_2, nK_2) = m + 2n - 1.$$

The lower bound for this Ramsey number is achieved by G_1 in Fig. 2.20. This graph is the red/blue coloring of K_{m+2n-2} formed by replacing one vertex in a red K_2 with a red K_{m-1} and the other vertex with a blue K_{2n-1}. Since every edge in a red matching must contain at least one vertex from the red K_{m-1}, there are not enough vertices to form a red mK_2. Every blue matching must be entirely contained in the blue K_{2n-1}, preventing a blue nK_2 from existing. Hence, the coloring given in Fig. 2.20 is a critical coloring for (mK_2, nK_2). When $n = m$, the colors can be switched and we obtain a second critical coloring for (mK_2, mK_2) as given by G_2 in Fig. 2.20. In the following theorem, it is shown that the two colorings in Fig. 2.20 are the only critical colorings for (mK_2, nK_2).

Theorem 2.32 ([47, 50]) *If $n \geq m \geq 1$ and G_1 and G_2 are the graphs given in Fig. 2.20, then*

$$\text{Crit}(mK_2, nK_2) = \begin{cases} \{G_1\} & \text{if } n > m \\ \{G_1, G_2\} & \text{if } n = m. \end{cases}$$

Proof Let $n \geq m \geq 1$ and proceed by (strong) induction on $m+n$. When $m = 1$ and $n \geq 1$, no critical coloring can contain red edges and we find that $r(K_2, nK_2) = 2n$. In other words, the only critical coloring is a blue K_{2n-1}, which corresponds with G_1 in Fig. 2.20 (and is a K_1 when $n = m = 1$). Now assume the theorem is true for all $n' \geq m' \geq 1$ such that $m + n > m' + n'$ and consider a critical coloring for (mK_2, nK_2), which is necessarily a red/blue coloring of K_{m+2n-2}.

Note that there must exist a vertex v incident with both red and blue edges, otherwise the graph must be monochromatic and hence, contains a red mK_2 or a blue nK_2. So, assume that a and b are vertices such that va is red and vb is blue. Let H denote the subgraph induced by $V(K_{m+2n-2}) - \{v, a, b\}$. The graph H is a red/blue coloring of K_{m+2n-5}, and since $r((m-1)K_2, (n-1)K_2) = m + 2n - 4$, it follows that H is a critical coloring for $((m-1)K_2, (n-1)K_2)$. By the inductive

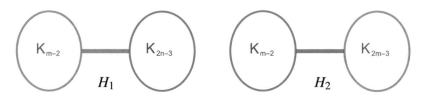

Fig. 2.21 The graph H_1 is the only critical coloring for $((m - 1)K_2, (n - 1)K_2)$ when $n > m$. When $n = m$, H_1 and H_2 are the only two critical colorings for $((m - 1)K_2, (n - 1)K_2)$

hypothesis, H is one of the colorings described in Fig. 2.21, which we label H_1 and H_2.

If $n > m$, then H is the coloring given by H_1, which is formed by replacing one vertex in a red K_2 with a red K_{m-2} and the other vertex with a blue K_{2n-3}. Let X denote the red K_{m-2} and Y denote the blue K_{2n-3}. There exists a blue matching of size $n - 2$ in Y. Suppose that the vertex a is joined to some vertex in X via a blue edge. Then a blue nK_2 is formed using this edge, the edge vb, and the blue matching in Y of size $n - 2$. So, a is adjacent to all vertices in X via red edges and $X \cup \{a\}$ is a red K_{m-1}, which we denote by X'. Now note that there exists a red matching of size $m - 2$ in which each edge contains one vertex from X and one vertex from Y. Also, such a matching can be formed using any $m - 2$ distinct vertices from Y. Suppose that b is joined to some vertex in Y by a red edge. Then a red mK_2 is formed using this edge, the edge va, and a red matching of size $m - 2$. So, b is adjacent to all vertices in Y by red edges and $Y \cup \{b\}$ is a red K_{2n-2}, which we denote by Y'.

Since Y' contains a blue matching of size $n - 1$, no blue edge can exist between v and the vertices in X' without forming a blue nK_2. So, all edges joining v to all vertices in X' must be red. There also exists a red matching of size $m - 1$ in which each edge has a single vertex from X' and a single vertex from Y'. Moreover, such a matching can be found having any selection of $m - 1$ vertices from X'. If v is adjacent to any vertex in Y' via a red edge, then a red mK_2 can be formed. It follows that $Y' \cup \{v\}$ is a blue K_{2n-1} and the original coloring of K_{m+2n-2} corresponds with G_1 in Fig. 2.20.

If $n = m$, then H could be given by H_1 or H_2. If it is H_1, then the above proof still holds. If H is given by H_2, then the switching the colors red and blue in the above proof above proof holds and the result is G_2 in Fig. 2.20. $\square$

Theorem 2.33 ([47, 50]) *If $n \geq m \geq 1$, then $r_*(mK_2, nK_2) = m$.*

Proof Starting with the graph G_1 in Fig. 2.20, add in a vertex v, connecting it by red edges to the red K_{m-1}. The result is a red/blue $K_{m+2n-2} \sqcup K_{1,m-1}$ that avoids a red mK_2 (since every edge in a red matching must include at least one vertex from the red K_{m-1}) and a blue nK_2. It follows that $r_*(mK_2, nK_2) \geq m$.

To prove the reverse inequality, consider a red/blue coloring of $K_{m+2n-2} \sqcup K_{1,m}$ and let v be the center vertex of the missing star. If a red mK_2 and blue nK_2 are to be avoided, then the red/blue coloring of the K_{m+2n-2} must be one of the critical colorings given in Fig. 2.20.

If $n > m$, then by Theorem 2.32 it must correspond with G_1. If any of the m edges joining v to the K_{m+2n-1} are blue, then a blue nK_2 is formed since the blue K_{2n-1} in G_1 contains a blue matching of size $n - 1$ using any choice of $2n - 2$ vertices. So, all m of the edges must be red and at least one of the red edges must join v to the blue K_{2n-1}. This red edge along with a red matching of size $m - 1$ in which each edge has one vertex in the red K_{m-1} and the other vertex is selected from the blue K_{2n-1} form a red mK_2. It follows that $r_*(mK_2, nK_2) \leq m$ whenever $n > m$.

If $n = m$, then by Theorem 2.32 the red/blue coloring of K_{3m-1} can correspond with either G_1 or G_2. If it corresponds with G_1, then the same proof as above applies. If it corresponds with G_2, then switching the colors in the above proof also holds. In both cases, it follows that $r_*(mK_2, mK_2) \leq m$. □

In 1975, Cockayne and Lorimer [24] proved that if $m_1, m_2, \ldots m_t \in \mathbb{N}$ and $m := \max\{m_1, m_2, \ldots, m_t\}$, then

$$r(m_1 K_2, m_2 K_2, \ldots, m_t K_2) = m + 1 + \sum_{i=1}^{t}(m_i - 1).$$

Generalizing Hook's result on matchings to the multicolor case, Xu et al. [92] proved the following theorem.

Theorem 2.34 ([92]) *If* $m_1 \geq m_2 \geq \cdots \geq m_t \geq 1$, *then*

$$r_*(m_1 K_2, m_2 K_2, \ldots, m_t K_2) = 1 + \sum_{i=2}^{t}(m_i - 1).$$

In [69], it was shown that for all $m \geq 2$ and $n \geq 1$, $r(K_m, nK_2) = m + 2n - 2$. When $m = 2$, the corresponding star-critical Ramsey number was determined in Theorem 2.33. In 2015, Li and Li [66] proved that (K_m, nK_2) is Ramsey-full for all $m \geq 3$.

Theorem 2.35 ([66]) *If* $m \geq 3$ *and* $n \geq 1$, *then* $r_*(K_m, nK_2) = m + 2n - 3$.

Proof When $n = 1$ the theorem follows from Theorem 1.1. So, assume that $m \geq 3$ and $n \geq 2$. It suffices to provide a red/blue coloring of $K_{m+2n-2} - e$ that avoids a red K_m and a blue nK_2. Start by replacing one vertex in a red K_2 with a red K_{m-2} and the other vertex with a blue K_{2n-1}. Introduce a vertex v and join it to all of the vertices in the blue K_{2n-1} with red edges and to exactly $m - 3$ of the vertices in the red K_{m-2} with red edges. One edge connecting v to the red K_{m-2} is the missing edge (e.g., see Fig. 2.22). The resulting $K_{m+2n-2} - e$ avoids a red K_m since the largest complete red subgraph includes at most one vertex from the blue K_{2n-1} or the vertex u, and hence, has order at most $m - 1$. A blue nK_2 is also avoided since the only blue edges are in the blue K_{2n-1}. It follows that (K_m, nK_2) is Ramsey-full, completing the proof of the theorem.

□

Fig. 2.22 A red/blue coloring of $K_{m+2n-2} - e$ that avoid a red K_m and a blue nK_2

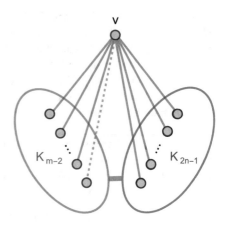

It is worth noting that in [66], Li and Li completely classified Crit(K_m, nK_2) for all $m \geq 2$ and $n \geq 1$. This classification was not necessary for the evaluation of $r_*(K_m, nK_2)$ since (K_m, nK_2) is Ramsey-full. However, the $m = 3$ case is needed later in the section on Fans (Sect. 2.6), where it is needed in in the proof of Theorem 2.41. So, a proof of this case is given below.

Theorem 2.36 ([66]) *For all $n \geq 1$, the critical colorings for (K_3, nK_2) are those with red subgraphs isomorphic to $K_{2i+1,2n-(2i+1)}$ and blue subgraphs isomorphic to $K_{2i+1} \cup K_{2n-(2i+1)}$, where $i \in \{0, 1, 2, \ldots, \lceil \frac{n}{2} \rceil\}$.*

Proof We proceed by induction on $n \geq 1$. When $n = 1$, a critical coloring for (K_3, K_2) cannot contain any blue edges, and hence, has at most 2 vertices. Thus, it must be a red K_2. Now assume the lemma is true for some $n - 1 \geq 1$ and consider a red/blue coloring of K_{2n} that avoids a red K_3 and a blue nK_2 (recall that $r(K_3, nK_2) = 2n + 1$). Such a coloring contains at least one blue edge, otherwise a red K_3 is formed since $n \geq 2$. Let uv be a blue edge. Removing uv produces a red/blue coloring of K_{2n-2}, which must be a critical coloring for $(K_3, (n - 1)K_2)$ since a blue $(n - 1)K_2$ and uv together would produce a blue nK_2. By the inductive hypothesis, this K_{2n-2} has red subgraph given by $K_{2i+1,2(n-1)-(2i+1)}$ and blue subgraph given by $K_{2i+1} \cup K_{2(n-1)-(2i+1)}$, for some $i \in \{0, 1, 2, \ldots, \lceil \frac{n-1}{2} \rceil\}$. Denote by X_1 the blue K_{2i+1} and by X_2 the blue $K_{2(n-1)-(2i+1)}$. If u joins to some vertex in X_1 and to some vertex in X_2 with red edges, then a red K_3 is formed. So, u must join to one of X_1 and X_2 via only blue edges. Without loss of generality, assume that u joins to X_2 via only blue edges. By a similar argument, v must join to all vertices in one of X_1 and X_2 via only blue edges. If v only joins to X_1 via blue edges, then $X_1 \cup \{v\}$ forms a blue K_{2i+2}, which contains a blue $(i + 1)K_2$. Likewise, $X_2 \cup \{u\}$ forms a blue $K_{2n-2i-2}$, which contains a blue $(n - i - 1)K_2$. These two blue graphs together produce a blue nK_2, leading to a contradiction. So, v must only join to vertices in X_2 via blue edges.

At this point, $X_2 \cup \{u, v\}$ forms a blue $K_{2n-2i-1}$, which contains a blue $(n - i - 1)K_2$. If ux is blue for any $x \in X_1$, then this edge, along with the remaining

$i K_2$ in X_1 and the $(n - i - 1)K_2$ in X_2 forms a blue nK_2. Thus, all edges joining u to X_1 must be red. The same is true for v and it follows that the critical coloring considered for (K_3, nK_2) must be one of those described in the statement of the theorem.   □

Now we turn our attention to the disjoint union of K_3. In 1975, Burr et al. [13] proved that if $n \geq m \geq 1$ and $n \geq 2$, then

$$r(mK_3, nK_3) = 2m + 3n.$$

In the following theorem, it will be shown that (mK_3, nK_3) is Ramsey-full.

Theorem 2.37 ([47, 50]) *If $n \geq m \geq 1$ and $n \geq 2$, then*

$$r_*(mK_3, nK_3) = 2m + 3n - 1.$$

Proof It suffices to provide a red/blue coloring of $K_{2m+3n} - e$ that avoids a red mK_3 and a blue nK_3. Begin by replacing one vertex in a red K_2 with a red K_{2m-1} and the other vertex with a blue K_{3n-1}. Introduce two vertices, called them u and v, joining them to the vertices in the red K_{2m-1} using blue edges and to the blue K_{3n-1} using red edges. The edge between u and v is the missing edge (see Fig. 2.23). The resulting $K_{2m+3n} - e$ lacks a red mK_3 since every red K_3 must contain at least two vertices from the red K_{2m-1}. It also lacks a blue nK_3 since every blue K_3 must be contained in the blue K_{3n-1}. It follows that $r_*(mK_3, nK_3) = 2m + 3n - 1$.   □

Lorimer and Mullins [70] proved that if $m \geq 1$ and $n \geq 2$, then

$$r(mK_3, nK_4) = \begin{cases} 2m + 4n + 1 & \text{if } n \geq m \\ 3m + 3n + 1 & \text{otherwise.} \end{cases}$$

Fig. 2.23 A red/blue coloring of $K_{2m+3n} - e$ that avoid a red mK_3 and a blue nK_3, where $n \geq m \geq 1$ and $n \geq 2$

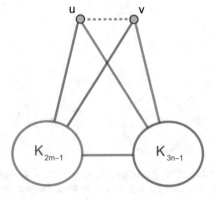

In the following theorem (first proved by Li and Li [66]), it is shown that (mK_3, nK_4) is Ramsey-full.

Theorem 2.38 ([66]) *If $m \geq 1$ and $n \geq 2$, then*

$$r_*(mK_3, nK_4) = \begin{cases} 2m + 4n \ if \ n \geq m \\ 3m + 3n \ otherwise. \end{cases}$$

Proof Let $p = r(mK_3, nK_4)$. Then the theorem will follow from constructing a red/blue coloring of $K_p - e$ that avoids a red mK_3 and a blue nK_4. First consider the case where $n \geq m$. When $n \geq m$, start with replacing one vertex in a red K_2 with a red K_{2m-1} and the other vertex with a blue K_{4n-1}. Introduce two vertices, call them u and v, connecting them to the red K_{2m-1} using blue edges and to the blue K_{4n-1} using red edges. The edge between u and v is the missing edge. The result (see the first image in Fig. 2.24) is a red/blue coloring of $K_{2m+4n} - e$. The only blue K_4-subgraphs must be entirely contained in the blue K_{4n-1}, which does not contain enough vertices to include n disjoint copies of K_4. For the red K_3 subgraphs, each one must contain at least two vertices from the red K_{2m-1}. Since there are fewer than $2m$ vertices in this complete subgraph, m disjoint copies of red K_3-subgraphs do not exist. Hence, (mK_3, nK_4) is Ramsey-full in this case.

When $n < m$, we use a similar construction, except that we start by replacing one vertex in a blue K_2 with a red K_{3m-1} and the other vertex with a blue K_{3n-1} (see the second image in Fig. 2.24). In this construction, the only red K_3-subgraphs must be entirely contained in the red K_{3m-1} and there are not enough vertices to have m disjoint copies of red K_3-subgraphs. Every blue K_4 must include at least 3 vertices from the blue K_{3n-1}. Having an nK_4 would then require there to be $3n$ such vertices, which we do not have. Thus, (mK_3, nK_4) is Ramsey-full in this case. □

In [10], Burr proved that if $k, \ell \geq 3$ are integers and m and n are large, then

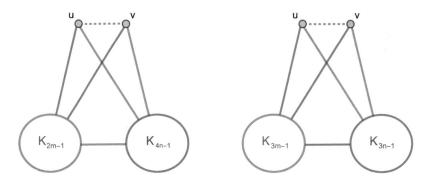

Fig. 2.24 Red/blue colorings of $K_p - e$, when $p = r(mK_3, nK_4)$. The first image corresponds with the case $n \geq m$ and the second image corresponds with the case $n < m$

$$r(mK_k, nK_\ell) = (k-1)m + (\ell - 1)n + \max\{m, n\} + r(K_{k-1}, K_{\ell-1}) - 2.$$

The following is Theorem 2 of [95], from which it follows that (mK_k, nK_ℓ) is Ramsey-full when m and n are large.

Theorem 2.39 ([95]) *Let $k, \ell \geq 3$ be integers. If m and n are large, then*

$$r_*(mK_k, nK_\ell) = (k-1)m + (\ell - 1)n + \max\{m, n\} + r(K_{k-1}, K_{\ell-1}) - 3.$$

Proof Without loss of generality, assume that $m \geq n$ and let

$$\begin{aligned}
p &= (k-1)m + (\ell - 1)n + \max\{m, n\} + r(K_{k-1}, K_{\ell-1}) - 2 \\
&= km + (\ell - 1)n + r(K_{k-1}, K_{\ell-1}) - 2.
\end{aligned}$$

Start with the disjoint union of a red K_{km-1}, a blue $K_{(\ell-1)n-1}$, and a red/blue critical coloring of $K_{r(K_{k-1}, K_{\ell-1})-1}$. Color all edges joining K_{km-1} to $K_{(\ell-1)n-1} \cup K_{r(K_{k-1}, K_{\ell-1})-1}$ blue and all edges joining $K_{(\ell-1)n-1}$ and $K_{r(K_{k-1}, K_{\ell-1})-1}$ red. No red mK_k exists as it would have to be entirely contained in the red K_{km-1}. Also, every blue K_ℓ must contain at least $\ell - 1$ vertices in the $K_{(\ell-1)n-1}$, preventing a blue nK_ℓ from being a subgraph.

Select a vertex a in the $K_{r(K_{k-1}, K_{\ell-1})-1}$ and introduce a new vertex b. Assign edge bx the same color as ax for any other vertex x and let a and b be the endpoints of the missing edge (see Fig. 2.25). No monochromatic complete graph with order greater than 1 can contain both a and b. By construction, the subgraph with a deleted is isomorphic to the corresponding subgraph with b deleted. So, we have constructed a red/blue coloring of $K_p - e$ that avoids a red mK_k and a blue nK_ℓ. It follows that (mK_k, nK_ℓ) is Ramsey-full.

Fig. 2.25 A red/blue coloring of $K_{r(mK_k, nK_\ell)} - e$ that avoids a red mK_k and a blue nK_ℓ

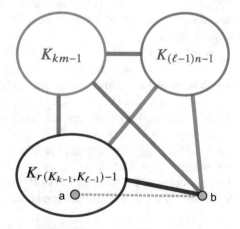

$\square$

2.6 Fans

In 2006, Salman and Broersma [81] proved that for all $m \geq 4$ and $m \geq 2n - 1 \geq 3$,

$$r(P_m, F_n) = 2m - 1.$$

In the process, they included the following lemma, which is needed for the evaluation of the corresponding star-critical Ramsey number.

Lemma 2.5 ([81]) *Let G be a graph with order $|V(G)| \geq n \geq 4$ that does not contain P_n as a subgraph. Suppose that the paths $P^1, P^2, \ldots, P^k$ are subgraphs of G chosen in the following way: $V(G) = \bigcup_{j=1}^{k} V(P^j)$, P^1 is a longest path in G, and for $k > 1$, P^{i+1} is a longest path in $V(G) - \bigcup_{j=1}^{i} V(P^j)$, where $1 \leq i \leq k-1$. Let u be a leaf of P^k. Then*

1. *$|V(P^1)| \geq |V(P^2)| \geq \cdots \geq |V(P^k)|$,*
2. *if $|V(P^k)| \geq \lfloor \frac{n}{2} \rfloor$, then $N_G(u) \subseteq V(P^k)$ and $N_G(u) \neq V(P^k)$,*
3. *if $|V(P^k)| \leq \lfloor \frac{n}{2} \rfloor - 1$, then $\deg_G(u) \leq \lfloor \frac{n}{2} \rfloor - 1$.*

Proof The first implication follows from the way in which the paths are chosen. This choice also leads to the observation that for any integer j with $1 \leq j < k$, the number of neighbors of u in $V(P^j)$ is at most

$$\left\lfloor \frac{|V(P^j)| + 1 - 2|V(P^k)|}{2} \right\rfloor$$

when $|V(P^j)| \geq 2|V(P^k)| + 1$, and is equal to 0 when $|V(P^j)| < 2|V(P^k)| + 1$. This is deduced by first ordering the neighbors of u on P^j according to their order of appearance on P^j, in a fixed orientation. Observe that between any two successive neighbors of u on P^j, there are at least $|V(P^k)|$ nonneighbors of u, from which the claim follows.

Assume that $|V(P^k)| \geq \lfloor \frac{n}{2} \rfloor$. Then $2|V(P^k)| + 1 \geq n > |V(P^1)|$ and it can be concluded that u does not have any neighbors in $V(G) - V(P^k)$. The second implication in the statement of the lemma follows.

Now assume that $|V(P^k)| < \lfloor \frac{n}{2} \rfloor$. If u does not have any neighbors in $V(G) - V(P^k)$, then the third implication of the lemma follows. So, assume that u does have some neighbors in $V(G) - V(P^k)$. Denote by $h_1, h_2, \ldots, h_t$ the numbers of vertices on $P^1, P^2, \ldots, P^k$ that contain a neighbor of u, chosen such that $h_t \geq \cdots \geq h_2 \geq h_1$, where $t \leq k$. Denote by $d_1, d_2, \ldots, d_t$ the numbers of neighbors of u on the corresponding paths. Then $h_1 = |V(P^k)| \geq d_1 + 1$ and $h_2 \geq 2h_1 + 2d_2 - 1$. Since u connects any two of the considered paths, if $t \geq 3$, we obtain

$$h_j \geq 2((h_{j-1}/2 + 2) + 2d_j - 1 = h_{j-1} + 2d_j + 2,$$

for $3 \leq j \leq t$. So, when $t \geq 2$, this implies that

$$h_t \geq 2(d_1 + d_2 + \cdots + d_t) + 2(t - 2) + 1 \geq 2|N_G(u)| + 1.$$

As $h_t \leq n - 1$ and $|N_G(u)|$ are both integers, the third implication in the lemma follows.   □

In 1973, Lawrence [58] proved that for all $m, n \in \mathbb{N}$ that satisfy $m \geq 2n$,

$$r(C_m, K_{1,n}) = m.$$

In 1974, Faudree et al. [31] proved that for all $m, n \in \mathbb{N}$ that satisfy $m \geq 2n \geq 4$,

$$r(P_m, C_{2n}) = m + n - 1.$$

Along with the above lemma, both of these results will be necessary in the evaluation of $r_*(P_m, F_n)$, under appropriate assumptions on the values of m and n.

Theorem 2.40 ([88]) *If $n \geq 2$ and $m \geq 2n + 1$, then $r_*(P_m, F_n) = m + n - 1$.*

Proof For the lower bound, begin by replacing the vertices in a blue K_2 with blue K_{m-1}-subgraphs. Add in vertex v joining it to all vertices in one of the red K_{m-1}-subgraphs using blue edges. Join v to the other K_{m-1} using exactly $n - 1$ blue edges and let the remaining $m - n$ edges be the missing edges. No red P_m exists since the largest connected red subgraph has order $m - 1$. No blue F_n exists since every blue K_3 must include vertex v along with one vertex from each of the red K_{m-1}-subgraphs. Only $n - 1$ such K_3-subgraphs can be formed that have a single vertex in common. From this construction, it follows that $r_*(P_m, F_n) \geq m + n - 1$.

To prove the reverse inequality, consider a red/blue coloring of $K_{2m-2} \sqcup K_{1,m+n-1}$ and let v be the center vertex for the missing star. Remove v and consider the resulting red/blue K_{2m-2}. If this coloring avoids a red P_m, then choose paths $P^1, P^2, \ldots, P^k$ in the underlying red subgraph G_R (i.e., the subgraph spanned by the red edges) as in Lemma 2.5 and let u be a leaf in P^k. Since the order of the G_R is $2m - 2$ and a red P_m is avoided, it follows that $k \geq 2$ and $|V(P^k)| \leq m - 1$. If $|V(P^k)| \leq \lfloor \frac{m}{2} \rfloor - 1$, then by Lemma 2.5, $\deg_{G_R}(u) \leq \lfloor \frac{m}{2} \rfloor - 1 \leq m - 2$. If $\lfloor \frac{m}{2} \rfloor \leq |V(P^k)| \leq m - 1$, then by Lemma 2.5, $\deg_{G_R}(u) \leq |V(P^k)| - 1 \leq m - 2$. It remains to be proven that there exists a blue F_n.

If $\deg_{G_R}(u) \leq \lfloor \frac{m}{2} \rfloor - 1$, then $\deg_{G_B}(u) \geq 2m - 3 - (\lfloor \frac{m}{2} \rfloor - 1) \geq m + n - 1$, where G_B is the subgraph spanned by the blue edges. Since $r(P_m, C_{2n}) = m + n - 1$ [31], $N_{G_B}(u)$ contains a blue C_{2n}, from which a blue F_n can be formed with u serving as its central vertex.

If $\deg_{G_R}(u) \geq \lfloor \frac{m}{2} \rfloor$, then by Lemma 2.5, it follows that $N_{G_R}(u) \subseteq V(P^k)$ and $N_{G_R}(u) \neq V(P^k)$. So, $|V(P^k)| \geq \lfloor \frac{m}{2} \rfloor + 1$. Since the order of G_R is $2m - 2$, $k = 2$ or $k = 3$. In the case $k = 3$, let $P^1 = x_1 x_2 \cdots x_{\ell_1}$ and $P^2 = y_1 y_2 \cdots y_{\ell_2}$, where $\ell_i = |V(P^i)|$. Since P^1 is the longest red path in G_R, $x_i y_i$ is blue for $i = 1, 2, \ldots, m$, producing a blue F_n with central vertex u.

Now consider the case $k = 2$. Since G_R does not contain a red P^m, $\ell_1 = \ell_2 = m - 1$ and the edges between P^1 and P^2 must all be blue. Reintroduce the vertex v (and the $m + n - 1$ edges joining it to the red/blue K_{2m-2}). If v joins to at least one vertex in P^1 and at least one vertex in P^2 via red edges, then a red P^m is formed. So, without loss of generality, assume that v is adjacent to P^2 via only blue edges. Since v is adjacent to at least m vertices in P^i for $i = 1, 2$, v must join to P^1 via at least one red edge, otherwise, there is a blue F_n with central vertex v. If $\{x_1, x_2, \ldots, x_{\ell_1}\}$ induces a red C_{m-1}, then including the vertex v, a red P_m is formed. So, assume that $\{x_1, x_2, \ldots, x_{\ell_1}\}$ does not induce a red C_{m-1}. Since $r(C_m, K_{1,n}) = m$ [58], $\{x_1, x_2, \ldots, x_{\ell_1}\}$ induces a blue $K_{1,n}$. Including any m vertices from $\{y_1, y_2, \ldots, y_{\ell_2}\}$ forms a blue F_n. Thus, it follows that $r_*(P_m, F_n) \leq m + n - 1$.   □

For $n \geq 2$, it was shown in [59] that $r(K_3, F_n) = 4n + 1$. The lower bound for this Ramsey number is achieved from a class of graphs, which we will denote by $\mathcal{G}$. In order to describe the graphs in $\mathcal{G}$, begin with a blue copy of $2K_{2n}$ and let M_k be the graph spanned by k independent edges, each of which joins vertices in different copies of K_{2n}. Here, we allow $0 \leq k \leq 2n$, with $k = 0$ corresponding with the empty set. Coloring the edges in M_k blue and all remaining edges red produces a red/blue coloring of K_{4n} with red subgraph isomorphic to $K_{2n,2n} - E(M_k)$ and blue subgraph containing all of the edges in $E(2K_{2n}) \cup E(M_k)$ (see Fig. 2.26). The red subgraph is bipartite, preventing the existence of a red K_3. Within the blue subgraph, observe that if the center vertex of a blue fan is contained in one of the copies of K_{2n}, then all of its vertices must be contained in the same copy of K_{2n} since the blue joining edges (when they exist) form a matching. As there are not enough vertices in a K_{2n} to contain an F_n, no blue F_n exists. In the following theorem, it is shown that the critical colorings for (K_3, F_n) are precisely those in $\mathcal{G}$.

Theorem 2.41 ([66]) *For all $n \geq 2$, the critical colorings for (K_3, F_n) are all in class $\mathcal{G}$.*

Proof Consider a critical coloring for (K_3, F_n), which necessarily is a red/blue coloring of K_{4n}. If the blue subgraph G_B contains a blue K_{2n}, then let ab be a red edge in which neither a nor b is contained in the blue K_{2n}. If any two edges from a to the blue K_{2n} are blue, then a blue F_n can be formed with ab being one of the edges in the matching nK_2 of $F_n = K_1 + nK_2$. The same is true for b, and it follows that each of a and b must be incident with at least $2n - 1$ red edges to the blue K_{2n}. Since $n \geq 2$, by the pigeonhole principle there exists some vertex c in the blue K_{2n} in which the subgraph induced by $\{a, b, c\}$ is a red K_3. Since this

Fig. 2.26 A red/blue coloring of K_{4n} that lacks a red K_3 and a blue F_n

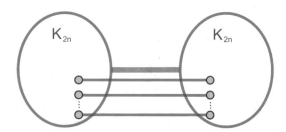

is avoided, we find that no such edge ab can be red, and G_B must actually contain a blue $2K_{2n}$. So, if we can argue that a critical coloring contains a blue K_{2n}, it will follow that it contains a blue $2K_{2n}$.

Let x denote a vertex in a critical red/blue coloring for (K_3, F_n) and consider the subgraph G'_B induced by $N_R(x)$ in G_B. Since no red K_3 exists, G'_B is a complete blue subgraph. In order to avoid a blue F_n, $\deg_R(x) \leq 2n$, from which it follows that $\deg_B(x) \geq 2n - 1$. On the other hand, $r(K_3, nK_2) = 2n + 1$ (e.g., see [69]) implies the upper bound to

$$2n - 1 \leq \deg_B(x) \leq 2n.$$

If $\delta(G_B) = 2n - 1$ and y is a vertex of blue degree $2n - 1$, then $|N_R(y)| = 2n$ and $r(K_2, K_{2n}) = 2n$ implies that there exists a blue K_{2n}.

Now assume that G_B is $2n$-regular (i.e., every vertex in G_B has degree equal to $2n$). For any vertex x, $\deg_B(x) = 2n$ and $r(K_3, nK_2) = 2n + 1$ implies that the subgraph induced by $N_B(x)$ is a critical coloring for (K_3, nK_2). By Theorem 2.36, its blue subgraph must be isomorphic to one of $K_{2i+1} \cup K_{2n-(2i+1)}$, for some $i \in \{0, 1, 2, \ldots, \lceil \frac{n}{2} \rceil\}$. If $n = 2$, then the blue subgraph induced by $N_B(x)$ is isomorphic to $K_1 \cup K_3$. We claim that for all $n \geq 3$, it is isomorphic to $K_1 \cup K_{2n-1}$. Suppose false, then it is $K_{2i+1} \cup K_{2n-(2i+1)}$ for some $i \in \{1, 2, \ldots, \lceil \frac{n}{2} \rceil\}$. Let X denote the blue K_{2i+1}, Y denote the blue $K_{2n-(2i+1)}$, and Z denote the subgraph induced by $N_R(x)$. Then Z is a blue K_{2n-1}. For any vertex $u \in X$, $\deg_B(u) = 2n$ implies that u has $2i + 1$ blue neighbors in $X \cup \{x\}$ and no neighbors in Y. So, u has exactly $2n - (2i + 1)$ blue neighbors in Z. Since

$$(2n - 1) - (2n - 2i - 1) = 2i \geq 2,$$

there is at least one vertex v in Z that is not a blue neighbor of u (see Fig. 2.27). Since $n \geq 3$, Y has at least 3 vertices, at most two of which are blue neighbors of v. Thus, there exists a vertex w in Y such that the subgraph induced by $\{u, v, w\}$ is a red K_3, giving a contradiction. So, the blue subgraph induced by $N_B(x)$ is a $K_1 \cup K_{2n-1}$, which along with x, contains a blue K_{2n}.

Fig. 2.27 A critical coloring for (K_3, F_n)

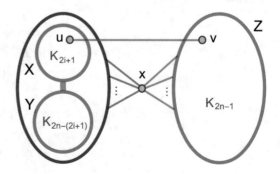

It has been shown that every member of $\mathcal{G}$ contains a blue subgraph isomorphic to K_{2n}, and hence, to $2K_{2n}$. So, the edge set for G_B has the form $E(2K_{2n}) \cup H$, where H consists of edges in a subgraph of $K_{2n,2n}$. Since no red K_3 or blue F_n exists, the maximum degree of any vertex in H is at most 1. The only possibilities are that H contains no edges or is isomorphic to M_k, for some $1 \le k \le 2n$.   □

Theorem 2.42 ([66]) *If $n \ge 2$, then $r_*(K_3, F_n) = 2n + 2$.*

Proof One can easily confirm that F_n is 3-good. So, Theorem 1.3 implies that $r_*(K_3, F_n) \ge 2n + 2$. We also provide an explicit construction demonstrating this bound. First, consider the graph in $\mathcal{G}$ that corresponds with $k = 0$. That is, consider the red/blue coloring of K_{4n} formed by replacing the vertices in a red K_2 with blue copies of K_{2n}. As we have already seen, this is a critical coloring for (K_3, F_n). Add in a vertex v and connect it to one copy of K_{2n} with all red edges, and the other copy with a single blue edge. The resulting red/blue coloring of $K_{4n} \sqcup K_{1,2n+1}$ avoids a red K_3 and a blue F_n (see Fig. 2.28).

To prove the reverse inequality, consider a red/blue coloring of $K_{4n} \sqcup K_{1,2n+2}$ and let v be the center vertex for the removed star. Upon deleting v, if the resulting K_{4n} lacks a red K_3 and a blue F_n, then it must be in $\mathcal{G}$ by Theorem 2.41. The blue subgraph of this K_{4n} contains two vertex-disjoint copies of blue K_{2n}-subgraphs, which we denote by X and Y. Denote their vertex sets by $\{x_1, x_2, \ldots, x_{2n}\}$ and $\{y_1, y_2, \ldots, y_{2n}\}$, respectively. If v is adjacent to any two vertices in X (or Y) via blue edges, then a blue F_n is formed. So, assume v is adjacent to at most one vertex in X and at most one vertex in Y via blue edges. Of the remaining $2n - 1$ vertices in X and $2n - 1$ vertices in Y, v connects to $2n$ of them. So, there exist at least two red edges between v and one of X and Y (say, X), and at least one red edge between v and Y. Without loss of generality, suppose that vx_1, vx_2, and vy_1 are red. Since the K_{4n} came from $\mathcal{G}$, it follows that at most one of x_1y_1 and x_2y_1 is blue. Without loss of generality, suppose that x_1y_1 is red. Then the subgraph induced by $\{v, x_1, y_1\}$ is a red K_3, from which it follows that $r_*(K_3, F_n) \le 2n + 2$.   □

Fig. 2.28 A red/blue coloring of $K_{4n} \sqcup K_{1,2n+1}$ that avoids a red K_3 and a blue F_n

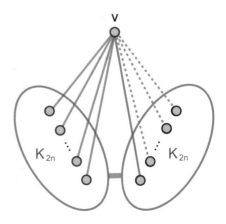

In [85], it was proved that for all $n \geq 4$,

$$r(K_4, F_n) = 6n + 1.$$

The following theorem includes a useful property possessed by the critical colorings for (K_4, F_n). Its proof is rather involved (following from Lemma 12 and Propositions 7, 9, and 10 of [40]), so we refrain from including it here.

Theorem 2.43 ([40]) *For all $n \geq 4$, every red/blue coloring of K_{6n} that avoids a red K_4 and a blue F_n necessarily contains a blue $3K_{2n}$.*

Theorem 2.44 ([40]) *If $n \geq 4$, then $r_*(K_4, F_n) = 4n + 2$.*

Proof To prove the lower bound for $r_*(K_4, F_n)$, we must provide a red blue coloring of $K_{6n} \sqcup K_{1,4n+1}$ that does not contain a red K_4 or a blue F_n. Begin with a red K_3 in which every vertex is replaced by a blue K_{2n}. Introduce a vertex v, joining it with red edges to all of the vertices in two of the copies of K_{2n}. Join v to the third copy of K_{2n} with a single blue edge to one of its vertices (see Fig. 2.29). The resulting red/blue coloring of $K_{6n} \sqcup K_{1,4n+1}$ avoids a red K_4 since such a subgraph would require a single vertex from each of the blue K_{2n}-subgraphs, along with v. It also avoids a blue F_n since v is not included in any such graph (it has blue degree 1) and there are not enough vertices in the blue K_{2n}-subgraphs to produce a blue F_n. It follows that $r_*(K_4, F_n) \geq 4n + 2$. Note that this lower abound also follows from observing that F_n is 4-good and applying Theorem 1.3.

To prove the reverse inequality, consider a red/blue coloring of $K_{6n} \sqcup K_{1,4n+2}$ and let w be the center vertex for its missing star (i.e., w is he vertex of degree $4n + 2$. Removing w results in a red/blue coloring of K_{6n} that, if a red K_4 and a blue F_n are avoided, must contain a blue $3K_2$ by Theorem 2.43. If a blue F_n is avoided, the only blue edges that can interconnect two distinct blue K_{2n}-subgraphs must form a matching. Denote by M_{k_i} the graph spanned by k_i independent edges, each of which spans a different pair of blue K_{2n} subgraphs, where the pair chosen corresponds with $i \in \{1, 2, 3\}$. Here, we allow $0 \leq k_i \leq 2n$ for each i. The colorings

Fig. 2.29 A red/blue coloring of $K_{6n} \sqcup K_{1,4n+1}$ that avoids a red K_4 and a blue F_n

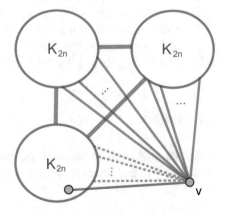

Fig. 2.30 A red/blue
coloring of $K_{6n} \sqcup K_{1,4n+2}$

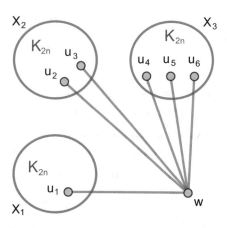

in $\text{Crit}(K_4, F_n)$ are precisely the colorings of K_{6n} in which the subgraphs spanned
by the blue edges have edge sets given by

$$E(3K_{2n}) \cup E(M_{k_1}) \cup E(M_{k_2}) \cup E(M_{k_3}).$$

The subgraphs spanned by their red edges are all subgraphs of the complete 3-partite
graph $K_{2n,2n,2n}$.

Denote the three blue K_{2n}-subgraphs by X_1, X_2, and X_3. In order to avoid a blue
F_n, w is joined by at most one blue edge to each X_i. Suppose that w is joined by j_i
red edges to X_i and assume that $j_1 \leq j_2 \leq j_3$. If $j_3 < 3$, then w is incident with at
most $9 < 4n + 2$ edges. If $j_2 < 2$, then w is incident with at most $2n + 4 < 4n + 2$
edges. If $j_1 < 1$, then w is incident with at most $4n + 1 < 4n + 2$ edges. In all
three cases, we reach a contradiction. Hence, $j_1 \geq 1$, $j_2 \geq 2$, and $j_3 \geq 3$. Denote
by $N_R^i(w)$ the red neighborhood of w in X_i. Let $u_1 \in N_R^1(w)$, $u_2, u_3 \in N_R^2(w)$, and
$u_4, u_5, u_6 \in N_R^3(w)$, where the six vertices listed are assumed to be distinct (see
Fig. 2.30). If a blue F_n is to be avoided, then u_1 joins to the vertices in $\{u_2, u_3\}$ using
at most one blue edge. Without loss of generality, assume that u_1u_2 is red. Similarly,
u_2 joins to the vertices in $\{u_4, u_5, u_6\}$ using at most one blue edge. Without loss
of generality, assume that u_2u_4 and u_2u_5 are red. Finally, u_1 joins to the vertices
in $\{u_4, u_5\}$ using at most one blue edge. Assuming that u_1u_4 is red, the subgraph
induced by $\{w, u_1, u_2, u_4\}$ is a red K_4. Therefore, $r_*(K_4, F_n) \leq 4n + 2$.   □

2.7 Books

Recall the definition of the book $B_n := K_2 + nK_1$. In the case where $n = 2$,
$B_2 = K_4 - e$ and Theorem 2.13 is equivalent to $r_*(P_4, B_2) = 4$. The following
theorem also involves the book B_2.

Fig. 2.31 A red/blue
coloring of $K_7 - E(K_{1,2})$ that
lacks a red B_2 and a blue K_3
(and hence, a blue $K_{1,3} + e$)

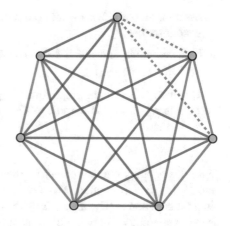

Theorem 2.45 ([5]) $r_*(B_2, K_3) = 5$.

Proof In [22], it was shown that $r(B_2, K_3) = 7$, from which it follows that B_2 is
3-good. Thus, Theorem 1.3 implies that $r_*(B_2, K_3) \geq 5$ (also, see Fig. 2.31). To
prove that $r_*(B_2, K_3) \leq 5$, we must argue that every 2-coloring of $K_7 - e$ contains
a red B_2 or a blue K_3. Consider an arbitrary red/blue coloring of the edges of $K_7 - e$
and denote by a and b the vertices of the missing edge. Removing vertex b produces
a 2-coloring of K_6, which must contain a red K_3 or a blue K_3 since $r(K_3, K_3) = 6$
[37]. In the latter case, we are done, so assume there is a red K_3.

Case 1 Suppose that a is not one of the vertices in the red K_3. Label the vertices
in the red K_3 by x, y, z and the other vertices a, b, c, d. If any of a, b, c, d
is adjacent via two or more red edges to $\{x, y, z\}$, then a red B_2 is formed.
Otherwise, each of a, b, c, d is adjacent via at least two blue edges to $\{x, y, z\}$.
If the subgraph induced by $\{a, b, c, d\}$ is not a red B_2, then at least one of its
edges is blue. Without loss of generality, suppose that ac is blue. Then a and
c are each adjacent via at least two blue edges to $\{x, y, z\}$. By the pigeonhole
principle, there is a vertex, say x, in which ax and cx are both blue, forcing the
subgraph induced by $\{a, c, x\}$ to be a blue K_3.

Case 2 Suppose that a is one of the vertices in the red K_3. Label the other two
vertices in the red K_3 by x and y and label the other vertices b, c, d, e. If any
of b, c, d, e is adjacent via two or more red edges to $\{a, x, y\}$, then a red B_2
is formed. Otherwise, each of c, d, e is adjacent via at least two blue edges
to $\{a, x, y\}$. If any edge in the subgraph induced by $\{c, d, e\}$ is blue, then the
pigeonhole principle can again be used to argue the existence of a blue K_3. So,
assume that the subgraph induced by $\{c, d, e\}$ is a red K_3. Then at least two of
the edges bc, bd, and be are blue (otherwise, the subgraph induced by $\{b, c, d, e\}$
contains a red B_2). Without loss of generality, suppose that bc and bd are blue. If
b is adjacent via red edges to both x and y, then a red B_2 is formed. So, assume
that one such edge, say bx, is blue. If either cx or dx is blue, then including b, we

obtain a blue K_3. If they are both red, then the subgraph induced by $\{x, c, d, e\}$ contains a red B_2.

In both cases, we find that a 2-colored $K_7 - e$ contains a red B_2 or a blue K_3. It follows that $r_*(B_2, K_3) \leq 5$.   □

Denote by $K_{1,3}+e$ the graph formed by adding an edge between two nonadjacent vertices in $K_{1,3}$. It is isomorphic to $K_3 \sqcup K_{1,1}$. Chvátal and Harary [22] proved that

$$r(B_2, K_{1,3} + e) = 7.$$

With this Ramsey number comes an interesting phenomenon: $K_{1,3} + e$ is B_2-good and B_2 is $(K_{1,3} + e)$-good. Theorem 1.3 implies that that $r_*(B_2, K_{1,3} + e) \geq 4$ in the first case and $r_*(B_2, K_{1,3} + e) \geq 5$ in the second case. The second implication is stronger and the following theorem shows this bound to be exact.

Theorem 2.46 ([5]) $r_*(B_2, K_{1,3} + e) = 5$.

Proof As mentioned above, $r_*(B_2, K_{1,3} + e) \geq 5$. This lower bound also follows from Fig. 2.31. Consider a red/blue coloring of $K_6 \sqcup K_{1,5}$ and denote the vertices of the missing edge by a and b. By Theorem 2.45, this coloring contains a red B_2 or a blue K_3. Assume the latter case, and consider two cases, based on whether or not the missing edge is incident with a vertex in the blue K_3. Note that it is not possible for both a and b to be in the blue K_3, since they are the vertices that make up the missing edge.

Case 1   Suppose that neither a nor b are contained in the blue K_3. Denote the vertices in the K_3 by x, y, and z and the other two vertices by c and d. If any edge joining $\{x, y, z\}$ to $\{a, b, c, d\}$ is blue, then a blue $K_{1,3}+e$ is formed. Assume that all such edges are red. If any edge (other than the missing edge) in the subgraph induced by $\{a, b, c, d\}$ is red, then a red B_2 is formed. Otherwise, all such edges are blue, and a blue $K_{1,3} + e$ is formed.

Case 2   Assume that one of a and b, say a, is in the blue K_3. Denote the vertices of the blue K_3 by a, x, and y and the other vertices by b, c, d, and e. Other than the missing edge, if any edge joining $\{a, x, y\}$ to $\{b, c, d, e\}$ is blue, then a blue $K_{1,3} + e$ is formed. So, assume that all such edges are red. If a red B_2 is to be avoided, then the subgraph induced by $\{b, c, d, e\}$ must acontain a blue $K_{1,3} + e$. In both cases, a red/blue coloring of $K_6 \sqcup K_{1,5}$ is shown to contain a red B_2 or a blue $K_{13} + e$.

It follows that $r_*(B_2, K_{1,3} + e) \leq 5$.   □

In 1978, Rousseau and Sheehan [80] proved that if $m, n \in \mathbb{N}$ satisfy $m > \frac{6n+7}{4}$, then $r(P_m, B_n) = 2m - 1$. Wang et al. [88] used this result in 2021 to prove the following theorem.

Theorem 2.47 ([88]) *If $n \geq 4$ and $m \geq 2n$, then $r_*(P_m, B_n) = m$.*

Proof For the lower bound, begin by replacing the vertices in a blue K_2 with copies of red K_{m-1}-subgraphs. Introduce a vertex v and join it to the red/blue K_{2m-2} using

$m - 1$ edges, all colored blue and connected to the same underlying red K_{m-1}. The resulting red/blue $K_{2m-2} \sqcup K_{1,m-1}$ avoids a red P_m since the largest red component has order $m - 1$. It also avoids a blue K_3, preventing the existence of a blue B_n. It follows that $r_*(P_m, B_n) \geq m$. Note that since P_m is B_n-good, this inequality also follows from Theorem 1.3.

To prove the upper bound, consider a red/blue coloring of $K_{2m-2} \sqcup K_{1,m}$ and let v denote the center vertex for the missing star. Removing v results in a red/blue coloring of K_{2m-2}, which we assume avoids a red P_m and a blue B_n. Let P denote a path of maximal length in the K_{2m-2}, with vertices given by $x_1 x_2 \cdots x_k$, where $2 \leq k \leq m - 1$. Let $X = \{x_1, x_2, \ldots, x_k\}$ and let Y consist of the vertices in the K_{2m-2} that are not in X. As P has maximal length, for every $u \in Y$, ux_1 and ux_k must be blue. Since $|Y| = 2m - 1 - k \geq m - 1 \geq n$, the edge $x_1 x_k$ must be red in order to avoid a blue B_n. So, $x_1 x_2 \cdots x_k x_1$ forms a red cycle C_k of length k. Since any two consecutive vertices in the C_k can be viewed as the end-vertices of a path of order k, all edges joining X to Y must be blue, otherwise a red path with length greater than k can be formed. In order to avoid a blue B_n, the subgraph induced by X must be a red K_k.

Suppose that $2 \leq k \leq n - 1$. Since $|Y| = 2m - k - 1$ and

$$r(P_{\lfloor (2m-k-1)/2 \rfloor}, B_{n-k}) \leq 2m - k - 2$$

(see [80]), Y contains a red $P_{\lfloor (2m-k-1)/2 \rfloor}$ or a blue B_{n-k}. In the latter case, the blue B_{n-k}, along with the vertices in X, forms a blue B_n, contradicting the assumption that the red/blue K_{2n-2} is a critical coloring for (P_m, B_n). In the former case, there exists a red $P_{\lfloor (2m-k-1)/2 \rfloor}$. Since

$$\left\lfloor \frac{2m - k - 1}{2} \right\rfloor \geq \frac{2m - k - 2}{2} \geq k + 1,$$

this red path has length longer than that of P, contradicting the assumption that P has maximal length. It follows that $|X| \geq n$, which forces Y is be a complete red graph in order to avoid a blue B_n. In order for Y to lack a red P_m, $|Y| \leq m - 1$. Then $|X| \leq m - 1$ implies that $|X| + |Y| = 2m - 2$ and $|X| = |Y| = m - 1$. The only critical coloring of K_{2m-2} with this structure must have red subgraph $2K_{m-1}$ and blue subgraph $K_{m-1,m-1}$.

Reintroducing v, and the m edges joining v to the K_{2m-2}, all such edges must be blue if a red P_m is to be avoided. Also, v must be adjacent to at least $\lceil \frac{m}{2} \rceil$ vertices in one of the red complete graphs and adjacent to at least one vertex in the other red complete graph. Since $\lceil \frac{m}{2} \rceil \geq n$ (it is assumed that $m \geq 2n$), there exists a blue B_n. It follows that $r_*(P_m, B_n) \leq m$. $\square$

For any $m, n \in \mathbb{N}$ satisfying $m \geq 3n - 3$ and any tree T_m of order m, Erdős et al. [29] proved that

$$r(T_m, B_n) = 2m - 1.$$

Building off of this Ramsey number, Wang et al. [87] proved that whenever $m \geq 3n$, $r_*(T_m, B_n) = m$. Their proof involved an algorithmic process and falls outside of the scope of this text. In the following theorem, this result is proved for the case where the tree being considered is a star.

Theorem 2.48 ([87])   *For all $m, n \in \mathbb{N}$ such that $m \geq 3n - 1$, $r_*(K_{1,m-1}, B_n) = m$.*

Proof   The lower bound follows from Theorem 1.5. Alternately, consider replacing the vertices in a blue K_2 with red K_{m-1}-subgraphs. Introduce a vertex v, joining it via blue edges to the vertices in one of the K_{m-1}-subgraphs. The edges joining the other K_{m-1}-subgraph are the missing edges, resulting in Fig. 2.32. The largest red component has order $m - 1$, from which it follows that a red $K_{1,m-1}$ is avoided. No blue K_3-subgraph exists, from which it follows that a blue B_n is avoided. This construction gives the lower bound $r_*(K_{1,m-1}, B_n) \geq m$.

To prove the reverse inequality, consider a red/blue coloring of $K_{2m-2} \sqcup K_{1,m}$ and let v be the center vertex of the missing star. Remove v and assume that the resulting K_{2m-2} avoids a red $K_{1,m-1}$ and a blue B_n. Denote by G this red/blue coloring of K_{2m-2} and let G_R and G_B be the subgraphs spanned by the red edges and blue edges, respectively. Then every vertex x in G satisfies

$$\deg_{G_B}(x) \geq 2m - 3 - (m - 2) = m - 1.$$

Select a vertex x. The remainder of the proof in separated into cases based on whether or not the subgraph of G_B induced by the neighborhood $N_{G_B}(x)$ contains a blue edge.

Case 1   Suppose that the subgraph of G_B induced by $N_{G_B}(x)$ contains a blue edge. Then G_B contains a blue K_3, whose vertex set we denote by $\{x, y, z\}$. Let

$$X := N_{G_B}(x) - \{y, z\}, \qquad Y := N_{G_B}(y) - (\{x\} \cup N_{G_B}(x)),$$

and

Fig. 2.32   A red/blue coloring of $K_{2n-2} \sqcup K_{1,m-1}$ that lacks a red $K_{1,m-1}$ and a blue B_n

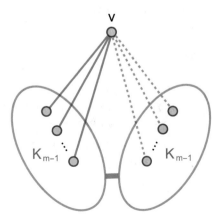

$$Z := N_{G_B}(z) - (N_{G_B}(x) \cup N_{G_B}(y)).$$

As every vertex in G_B has degree at least $m - 1$, it follows that $|X| \geq m - 3$. If a blue B_n is avoided, then $|Y| \geq (m - 3) - (n - 2) = m - n - 1$, and $|Z| \geq (m - 3) - 2(n - 2) = m - 2n + 1$. Then the order of G satisfies

$$\begin{aligned} |V(G)| = 2m - 2 &\geq |\{x, y, z\} \cup X \cup Y \cup Z| \\ &\geq 3 + (m - 3) + (m - n - 1) + (m - 2n + 1) \\ &\geq 3m - 3n, \end{aligned}$$

which is equivalent to $m \leq 3n - 2$. This contradicts the assumption that $m \geq 3n - 1$, preventing this case from occurring.

Case 2 Suppose that the subgraph of G_B induced by $N_{G_B}(x)$ is a red complete graph. Since $|N_{G_B}(x)| = \deg_{G_B}(x) \geq m - 1$, if no red $K_{1,m-1}$ is contained in G, then $|N_{G_B}(x)| = m - 1$. Now choose a vertex $y \in N_{G_B}(x)$ and consider the neighborhood $N_{G_B}(y)$. Then $|N_{G_B}(y)| = \deg_{G_B}(y) \geq m-1$, and if the subgraph induced by $N_{G_B}(y)$ contains a blue edge, the proof follows the argument given in Case 1. So, the subgraph induced by $N_{G_B}(y)$ is a red complete graph. If a red $K_{1,m-1}$ is avoided, then $|N_{G_B}(y)| = m - 1$. It follows that $G_R = 2K_{m-1}$ and $G_B = K_{m-1,m-1}$.

If any red edge joins v to G, then a red $K_{1,m}$ is formed with that edge and the red K_{m-1} that joins to v. So all m edges joining v to G must be blue. At least $\lceil \frac{m}{2} \rceil$ vertices in one red K_{m-1} must join to v via blue edges and at least one vertex in the other red K_{m-1} must join to v via a blue edge. Since $\lceil \frac{m}{2} \rceil \geq n$ follows from the assumption that $m \geq 3n - 1$, a blue B_n is formed.

It follows that $r_*(K_{1,m-1}, B_n) \leq m$. $\qquad\square$

This section is concluded with two theorems about the general behavior of certain star-critical Ramsey numbers involving books.

Theorem 2.49 ([64]) *For all sufficiently large n,*

$$r_*(C_m, B_n) = \begin{cases} 2m & \text{if } \frac{9n}{10} \leq m < n \\ 2n + 1 & \text{if } n + 1 \leq m \leq \frac{10n}{9}. \end{cases}$$

Theorem 2.50 ([61]) *There exists a constant $c > 0$ such that*

$$r_*(B_n, B_n) \geq r(B_n, B_n) - 1 - n + c \log n \log \log n$$

for infinitely-many values of n.

2.8 Generalized Fans and Books

In this section, generalizations of the fan $F_n = K_1 + nK_2$ and book $B_n = K_2 + nK_1$ are considered. Recalling that the graph $K_{1,3} + e$ is formed by adding an edge between two nonadjacent vertices in $K_{1,3}$, it can also be described by $K_1 + (K_2 \cup K_1)$. In this sense, $K_{1,3} + e$ can be considered a generalized fan. The following is Theorem 3.4 of [5].

Theorem 2.51 ([5]) $r_*(K_{1,3} + e, K_3) = 4$.

Proof It was shown in [22] that $r(K_{1,3} + e, K_3) = 7$. It follows that $K_{1,3} + e$ is 3-good and Theorem 1.3 implies that $r_*(K_{1,3} + e, K_3) \geq 4$ (also, see Fig. 2.5). It remains to be shown that every red/blue coloring of the edges of $K_6 \sqcup K_{1,4}$ contains a red $K_{1,3} + e$ or a blue K_3.

Consider a red/blue coloring of $K_6 \sqcup K_{1,4}$ and denote by a the vertex that is incident with the two missing edges. Removing a produces a red/blue coloring of K_6, which necessarily contains a monochromatic K_3 (since $r(K_3, K_3,) = 6$). If there is a blue K_3, we are done, so assume there exists a red K_3 with vertices b, c, and d. Label the three remaining vertices e, f, and g. If any edge joining the sets $\{b, c, d\}$ and $\{a, e, f, g\}$ is red, then a red $K_{1,3} + e$ is formed. So, assume that all such edges are blue. Figure 2.33 shows the three cases resulting from the possible locations of the red K_3 relative to the missing $K_{1,2}$.

In all three cases, observe that if any edge in the subgraph induced by $\{a, e, f, g\}$ is blue, then a blue K_3 is formed by using that edge and the edges connecting it to vertex d. Otherwise, all such edges are red (other than those included in the missing $K_{1,2}$) and a red $K_{1,3} + e$ is formed. It follows that $r_*(K_{1,3} + e, K_3) \leq 4$. □

For all $n \geq 1$ and $t \geq 2$, define the *generalized fan* $F_{t,n} := K_1 + nK_t$. The case $t = 2$ corresponds with the usual fan F_n. In [42], Hao and Lin proved that $r(K_3, F_{3,n}) = 6n + 1$ and determined the corresponding star-critical Ramsey number. The following theorem will be necessary for the evaluation of $r_*(K_3, F_{3,n})$.

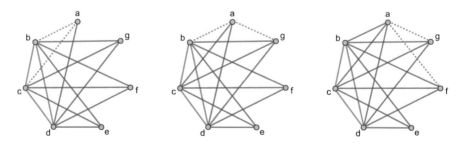

Fig. 2.33 Three cases showing the location of a red K_3 relative to a missing $K_{1,2}$ in a red/blue coloring of $K_6 \sqcup K_{1,4}$

Theorem 2.52 ([42]) *For all* $n \geq 4$, *the critical colorings for* $(K_3, F_{3,n})$ *are red/blue colorings of* K_{6n} *in which the subgraph spanned by the blue edges contains* $2K_{3n}$ *as a subgraph.*

Proof Consider a critical coloring of $(K_3, F_{3,n})$ (i.e., a red/blue coloring of K_{6n} that avoids a red K_3 and a blue $F_{3,n}$) with vertex set V. If any vertex $u \in V$ has blue degree $\deg_B(u) \geq 3n + 2$, then $r(K_3, nK_3) = 3n + 2$ [13] implies that there is a red K_3 or a blue $F_{3,n}$ (formed by the join of u and the blue nK_3), contradicting the assumption that we are considering a critical coloring. So, $\deg_B(u) \leq 3n + 1$ for all u, which implies that the red degree of every vertex $u \in V$ satisfies $\deg_R(u) \geq 3n - 2$. Since no red K_3 exists in this coloring, the red neighborhood $N_R(u)$ induces a complete blue subgraph of order at least $3n - 2$, which we call X_1. Now, let v be a vertex in X_1. Since $\deg_R(v) \geq 3n - 2$, and $N_R(v)$ does not contain any of the vertices in X_1, the neighborhood $N_R(v)$ induces a complete blue subgraph of order at least $3n - 2$, which we denote by X_2. Thus, the blue graph G_B contains a subgraph isomorphic to $2K_{3n-2}$.

Let $\{w, x, y, z\} \subseteq V - (X_1 \cup X_2)$. We claim that every vertex in $\{w, x, y, z\}$ is blue-adjacent to every vertex in either X_1 or X_2. Suppose false, then assume without loss of generality that w is adjacent to $u \in X_1$ and $v \in X_2$ via red edges. If $u_1, u_2 \in X_1$ are also joined to w via red edges, then vu, vu_1, and vu_2 cannot all be blue (otherwise, $X_2 \cup \{u, u_1, u_2\}$ contains a blue $F_{3,n}$). At least one of them must be red, say vu, and the subgraph induced by $\{u, v, w\}$ is a red K_3. Thus, w is adjacent to at most two vertices in X_1 via red edges. So, w is joined to at least $3n - 4$ vertices in X_1 using blue edges. The same is true for X_2. So, the blue neighborhood of w in X_1 contains a blue $(n - 2)K_3$ and the blue neighborhood of w in X_2 contains two disjoint triangles, together forming a blue $F_{3,n}$. This contradiction implies that w must join to either X_1 or X_2 using only blue edges, and the claim is proved.

So, every vertex in $\{w, x, y, z\}$ is blue-adjacent to every vertex in either X_1 or X_2. If at least three of them are blue-adjacent to all vertices in the same X_i, then a blue $F_{3,n}$ is produced. Now, assume that two of the vertices, say w and x, are blue-adjacent to all vertices in X_1 and y and z are blue-adjacent to all vertices in X_2. If w joins to at least three vertices in X_2 via blue edges, then these three vertices, along with w and X_2 form a blue $F_{3,n}$. So, w joins to at most two vertices in X_2 via blue edges. The same is true for x, and it follows that w and x must have a common neighbor via red edges in X_2. If wx is red, then a red K_3 is produced. Otherwise, wx is blue and $X_1 \cup \{w, x\}$ forms a blue K_{3n}. A similar argument implies that $X_2 \cup \{y, z\}$ forms a blue K_{3n}. $\square$

Theorem 2.53 ([42]) *For all* $n \geq 4$, $r_*(K_3, F_{3,n}) = 3n + 3$.

Proof Observe that since $r(K_3, F_{3,n}) = 6n + 1$, the fan $F_{3,n}$ is 3-good, and Theorem 1.3 implies that $r_*(K_3, F_{3,n}) \geq 3n + 3$. This lower bound can also be achieved by replacing the vertices in a red K_2 with blue K_{3n}-subgraphs. A vertex v can then join to all vertices in one K_{3n} via red edges and exactly two vertices in the other K_{3n} via blue edges. The remaining $3n - 2$ edges joining v to the second copy

Fig. 2.34 A red/blue coloring of $K_{6n} \sqcup K_{1,3n+2}$ that avoids a red K_3 and a blue $F_{3,n}$

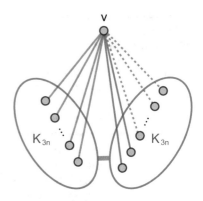

of K_{3n} are the missing edges and a red/blue $K_{6n} \sqcup K_{1,3n+2}$ is formed that lacks a red K_3 and a blue $F_{3,n}$ (see Fig. 2.34).

To prove the reverse inequality, consider a red/blue coloring of $K_{6n} \sqcup K_{1,3n+3}$ where v is the center vertex for the missing star. Removing v produces a red/blue K_{6n}. If a red K_3 and a blue $F_{3,n}$ are to be avoided, then by Theorem 2.52, it must contain $2K_{3n}$ as a blue subgraph. Denote the two blue K_{3n}-subgraphs by X_1 and X_2. Since v is incident with $3n + 3$ edges, it must be adjacent to at least three vertices in each of X_1 and X_2. Also, note that v must be adjacent to at least five vertices in one of X_1 and X_2. Without loss of generality, assume that v is adjacent to at least five vertices in X_1.

Let x_1, x_2, x_3, x_4, and x_5 be vertices in X_1 and y_1, y_2, and y_3 be vertices in X_2, all eight of which are adjacent to v. If v has at least three blue neighbors in X_1, then a blue $F_{3,n}$ is formed, so v joins to vertices in X_1 via at most two blue edges. Without loss of generality, assume that vx_1, vx_2, and vx_3 are red. Similarly, v joins to at most two vertices in X_2 via blue edges. Without loss of generality, assume that vy_1 is red. Now y_1 joins to X_1 using at most two blue edges. It follows that one of $x_1 y_1, x_2 y_1$, or $x_3 y_1$ must be red, forming a red K_3. It follows that $r_*(K_3, F_{3,n}) \leq 3n + 3$.   □

In [43], Hao and Lin proved the following result for sufficiently large n, which agrees with the $m = t = 3$ case of the star-critical Ramsey number proved in Theorem 2.53.

Theorem 2.54 ([43]) *For any fixed integers $m \geq 3$ and $t \geq 2$, there exists an $n_0 \in \mathbb{N}$ such that*

$$r_*(K_m, F_{t,n}) = (m - 2)tn + t,$$

for all $n \geq n_0$.

Further generalizing the concept of a generalized fan, for a graph H and $n \geq 2$, let $F_{H,n} := K_1 + nH$. Then $F_{K_t,n} = F_{t,n}$. Hamm et al. [41] proved the following theorem.

Theorem 2.55 ([41]) *Suppose that H is a graph of order h. Let $m \geq 3$, $n \geq 1$, and t satisfy*

$$ht \geq \max \left\{ (m+1)(m+C(H,m)+h)+3, \frac{(m-1)^2}{m-2}((n-1)(m-1)+h) \right\},$$

where $C(H,m)$ is a constant depending only on H and m. Then

$$r_*(F_{H,t,}, nK_m) = ht(m-2) + n + \delta(H).$$

Define the *generalized book* $B_p(n) := K_p + nK_1$. Such books were considered in Hao and Lin's 2018 paper [43]. They proved the following asymptotic result.

Theorem 2.56 ([43]) *For fixed $m \geq 3$ and $p \geq 2$,*

$$r_*(K_m, B_p(n)) = (m - 2 + o(1))n,$$

as $n \rightarrow \infty$.

Li et al. [62] considered the case of a cycle versus a generalized book and proved the following theorem, for values of n sufficiently large.

Theorem 2.57 ([62]) *Let $k, p \in \mathbb{N}$. If n is large, then*

$$r_*(C_{2k+1}, B_p(n)) = r(C_{2k+1}, B_p(n)) - n.$$

Additional results of this nature, describing the asymptotic behavior of star-critical Ramsey numbers involving generalizations of fans and books were further considered in a sequence of papers by Li et al. (see [62, 63], and [65]). We conclude this section by stating their main results.

Theorem 2.58 ([62]) *Let G be a graph and $k \in \mathbb{N}$. If n is large, then*

$$r_*(C_{2k+1}, K_1 + nG) = r(C_{2k+1}, K_1 + nG) - |V(G)|n + \delta(G).$$

Theorem 2.59 ([63]) *For graphs G_1 and G_2, if n is large, then*

$$r_*(K_2 + G_1, K_1 + nG_2) = r(K_2 + G_1, K_1 + nG_2) - |V(G_2)|n + \delta(G_2).$$

Theorem 2.60 ([65]) *Suppose that $m \in \mathbb{N}$ is fixed. Then for all graphs G,*

$$r_*(F_m, K_1 + nG) = r(F_m, K_1 + nG) - 1 - (1 + o(1))|V(G)|n,$$

as $n \rightarrow \infty$.

Recall that $K_k(m)$ denotes the complete k-partite graph with m vertices in each partite set.

Theorem 2.61 ([65]) *Let $k \geq 2$ and $m \geq 1$ be fixed integers. For a given graph G, if m is odd, or if m is even and $|V(G)|n$ is odd, then*

$$r_*(K_1 + K_k(m), K_1 + nG) = r(K_1 + K_k(m), K_1 + nG) - 1 - (1 + o(1))|V(G)|n,$$

as $n \to \infty$.

Chapter 3
Generalizations of Star-Critical Ramsey Numbers

3.1 Star-Critical Gallai-Ramsey Numbers

Motivated by Gallai's foundational work[1] on transitively orientated graphs [33],
Gyárfás and Simonyi [39] defined a *Gallai t-coloring* of a graph G to be a map
$c : E(G) \longrightarrow \{1, 2, \ldots, t\}$ that lacks rainbow triangles. That is,

$$|\{c(xy), c(yz), c(xz)\}| \leq 2$$

for all $x, y, z \in V(G)$. Observe that when $t \leq 2$, every t-coloring is a Gallai
t-coloring. The *Gallai-Ramsey number* $gr(G_1, G_2, \ldots, G_t)$ is defined to be the
least natural number p such that every Gallai t-coloring of K_p contains a subgraph
isomorphic to G_i in color i, for some $1 \leq i \leq t$. The reader interested in learning
about Gallai-Ramsey theory is encouraged to consult the text by Magnant and Salehi
Nowbandegani [71]. The following structural result for Gallai colorings can be
found in [39] and is a reinterpretation of Gallai's original result in [33].

Theorem 3.1 ([39]) *Every Gallai-colored complete graph can be obtained by substituting Gallai-colored complete graphs into the vertices of a 2-colored complete graph of order at least* 2.

In this theorem, the 2-colored complete graph is called the *base graph* of the
Gallai coloring and the Gallai-colored complete graphs that are substituted into each
vertex in the base graph are called the *blocks* of the Gallai coloring. One often
chooses the base graph B of a Gallai-colored complete graph to have minimal order
n. In this case, note that $n \neq 3$ since one of the vertices in B is forced to be incident
with edges using only one color. Call this vertex x. One can then form a Gallai

---

[1] An English translation of Gallai's paper [33] can be found in the book *Perfect Graphs*, by
Ramírez-Alfonsín and Reed [78].

© The Author(s), under exclusive license to Springer Nature Switzerland AG 2023
M. R. Budden, *Star-Critical Ramsey Numbers for Graphs*, SpringerBriefs in
Mathematics, https://doi.org/10.1007/978-3-031-29981-0_3

partition with a base graph isomorphic to K_2 by viewing the subgraph joining the other two vertices of B into a single vertex.

In 2022, Su and Liu [84] extended the star-critical concept to Gallai-Ramsey theory. The *star-citical Gallai-Ramsey number* $gr_*(G_1, G_2, \ldots, G_t)$ is the least natural number k such that every Gallai t-coloring of $K_{gr(G_1,G_2,\ldots,G_t)-1} \sqcup K_{1,k}$ contains a copy of G_i in color i, for some $1 \le i \le t$. When $G_1 = G_2 = \cdots = G_t$, we write $gr_*^t(G_1)$ for the corresponding t-color Gallai-Ramsey number. Observe that when $t \le 2$, the star-critical Gallai-Ramsey number corresponds with the usual star-critical Ramsey number. In the special case where

$$gr_*(G_1, G_2, \ldots, G_t) = gr(G_1, G_2, \ldots, G_t) - 1,$$

we say that $(G_1, G_2, \ldots, G_t)$ is *Gallai-Ramsey-full*. When $G_1 = G_2 = \cdots = G_t$, we say that G_1 is Gallai-Ramsey-full.

The proof of Theorem 1.1 easily extends to Gallai-Ramsey numbers, leading to the observation that collections of complete graphs are Gallai-Ramsey-full (cf. Theorem 1 in [84]). That is, for all $n_i \ge 2$ and $1 \le i \le t$,

$$gr_*(K_{n_1}, K_{n_2}, \ldots, K_{n_t}) = gr(K_{n_1}, K_{n_2}, \ldots, K_{n_t}) - 1.$$

This result and others led Su and Liu [84] to state the following conjecture (see Open Problem 5 in Sect. 4.1).

Conjecture 3.1 ([84]) If G is a graph that does not contain any isolated vertices, then G is Ramsey-full if and only if G is Gallai-Ramsey-full.

In 2010, Faudree et al. [30] proved that for all $t \ge 2$,

$$gr^t(C_4) = t + 4.$$

Su and Liu [84] proved that C_4 is Gallai-Ramsey-full.

Theorem 3.2 ([84]) *For all $t \ge 2$, $gr_*^t(C_4) = t + 3$.*

Proof This theorem will follow from providing a Gallai t-coloring of $K_{t+4} - e$ that avoids a monochromatic copy of C_4. For $t = 2$, begin by partitioning the edge set of K_5 into two edge-disjoint cycles of length 5. Label its vertices $x_1, x_2, \ldots, x_5$ and suppose that $x_1x_2x_3x_4x_5x_1$ is a C_5 in color 1 and $x_1x_3x_5x_2x_4x_1$ is a C_5 in color 2. Label this Gallai 2-coloring of K_5 as G_2. For $t \ge 3$, G_t is formed by adding a vertex x_{t+3} to G_{t-1} and assigning all edges joining x_{t+3} to G_{t-1} the color t. For example, the coloring G_4 is shown in Fig. 3.1 with colors 1, 2, 3, and 4 given as red, blue, green, and orange, respectively.

Given G_t, introduce a vertex v. Color edges vx_2 and vx_3 with color 1 and edges vx_1 and vx_4 with color 2. Edge vx_5 will be the missing edge. For $6 \le i \le t + 3$, color edge vx_i with color $i - 3$ (i.e., the same color that edges x_ix_j received, where $j < i$). This construction avoids rainbow triangles. The subgraphs of the resulting $K_{t+4} - e$ in colors 1 and 2 are both isomorphic to C_5 with vertex v joining to two

Fig. 3.1 A Gallai 4-coloring
of K_7 that avoids a
monochromatic C_4

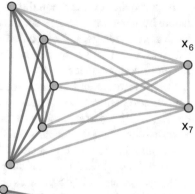

Fig. 3.2 A Gallai 4-coloring
of K_6 that avoids a
monochromatic P_4

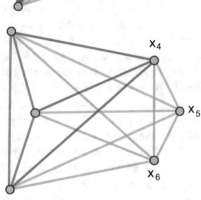

adjacent vertices in the cycle, while the subgraphs in colors $3, 4, \ldots, t$ are all stars. Hence, a monochromatic C_4 is avoided and it follows that C_4 is Gallai-Ramsey-full.
□

Faudree et al. [30] also proved that for all $t \in \mathbb{N}$,

$$gr^t(P_4) = t + 3.$$

Su and Liu [84] considered the corresponding star-critical Gallai-Ramsey number.

Theorem 3.3 ([84]) *For any $t \in \mathbb{N}$, $gr_*^t(P_4) = t$.*

Proof Begin by constructing a Gallai t-coloring of K_{t+2} that avoids a monochromatic P_4. Let x_1, x_2, x_3 be the vertices of a K_3 in color 1. Label this coloring G_1. For $t \geq 2$, form G_t by adding a vertex x_{t+2} to G_{t-1} and assigning all edges joining x_{t+2} to G_{t-1} the color t. For example, the coloring G_4 is shown in Fig. 3.2 with colors 1, 2, 3, and 4 given as red, blue, green, and orange, respectively.

Given G_t, introduce a vertex v. For each $4 \leq i \leq t+2$, color edge vx_i using color $i - 2$ and let $vx_1, vx_2,$ and vx_3 be the missing edges. The resulting $K_{t+2} \sqcup K_{1,t-1}$ avoids rainbow triangles. Its subgraph spanned by edges in color 1 is a K_3, while its subgraphs spanned by edges in colors $2, 3, \ldots, t$ are all stars. So, a monochromatic P_4 is also avoided. It follows that $gr_*^t(P_4) \geq t$.

To prove the reverse inequality, consider a Gallai t-coloring of $K_{t+2} \sqcup K_{1,t}$ and denote by v the center vertex for the missing star. If $t = 1$, then the monochromatic $K_3 \sqcup K_{1,1}$ contains a P_4 in color 1. If $t = 2$, then the theorem follows from Theorem 2.2, where it was shown that $r_*(P_4, P_4) = 2$. So, for the remainder of this proof, assume that $t \geq 3$.

Delete vertex v and assume that the resulting Gallai t-coloring of K_{t+2} avoids a monochromatic P_4. By Theorem 3.1, the vertices can be partitioned into sets $X_1, X_2, \ldots, X_q$ corresponding to the vertices in its 2-colored base graph. Furthermore, assume that $q \geq 2$ is minimal with regard to this property and note that $q \neq 3$, as was noted earlier in this section. Since $t + 2 \geq 5$ vertices are contained in this complete graph, the Ramsey number $r(P_4, P_4) = 5$ [34] implies that $q \leq 4$.

If $q = 4$, then $t+2 \geq 5$ vertices are partitioned into the four sets X_1, X_2, X_3, X_4, from which it follows that at least one of the sets, say X_1 has cardinality at least 2. If any other set X_i, where $2 \leq i \leq 4$, has cardinality at least 2, then a monochromatic P_4 can be formed using the edges joining X_1 and X_i. The only other possibility is that $|X_1| = t - 1$ and $|X_2| = |X_3| = |X_4| = 1$. Let $x, y \in X_1$ and $z_i \in X_i$ for $2 \leq i \leq 4$. By the pigeonhole principle, at least two of z_2, z_3, z_4 join to X_1 with the same color edges. Without loss of generality, assume that z_2 and z_3 both join to X_1 with the same color edges. Then xz_2yz_3 is a monochromatic P_4.

Finally, assume that $q = 2$ and note that $|X_1| = t + 1$ and $|X_2| = 1$. If $t = 3$, then a monochromatic P_4 is formed if the color of the edges that join X_1 and X_2 appears on any edge in the subgraph spanned by X_1. So, this subgraph must be a 2-colored K_4. In order for it to avoid a monochromatic P_4, the subgraph spanned by edges in one color must be a $K_{1,3}$ while the subgraph spanned by the other color must be a K_3 (that is, it must be the coloring labeled G_2). Now assume that for $t \geq 4$, every Gallai $(t - 1)$-coloring of K_{t+1} with base graph isomorphic to K_2 that avoids a monochromatic P_4 is given by G_{t-1} and contains a monochromatic triangle. Consider a Gallai t-coloring of K_{t+2} with base graph isomorphic to K_2 that avoids a monochromatic P_4. If the color that joins X_1 and X_2 appears on any edge in the subgraph induced by X_1, then a P_4 is formed in that color. Thus, X_1 must be a Gallai $(t - 1)$-coloring of K_{t+1}, which contains a monochromatic K_3 by the inductive hypothesis. It also follows that the coloring is given by G_t.

So, the critical Gallai t-coloring of K_{t+2} is given by G_t with vertices given by $x_1, x_2, \ldots, x_{t+2}$ as described at the beginning of this proof. Reintroducing the vertex v, and the t edges joining v to the Gallai t-coloring of K_{t+2}, note that at least one such edge must join to $\{x_1, x_2, x_3\}$ (the K_3 in color 1). Suppose that edge is vx_1. If it receives color 1, then a monochromatic P_4 is formed, so assume it receives some color i, where $2 \leq i \leq t$. Then $vx_1x_{i+2}x_2$ is a P_4 in color i. In all cases, a monochromatic P_4 is formed, from which it follows that $gr_*^t(P_4) \leq t$.   □

Theorem 3.4 ([84]) *For all $t \geq 3$,*

$$gr_*^t(K_{1,m}) = \begin{cases} 2m - 2 & \text{if } m \geq 12 \text{ is even} \\ m & \text{if } m \geq 3 \text{ is odd.} \end{cases}$$

Interestingly, the value of the star-critical Gallai-Ramsey number in Theorem 3.4 does not depend upon the value of t. Su and Liu [84] also considered how other star-critical Gallai-Ramsey numbers grow relative to t, resulting in the following theorem.

Theorem 3.5 ([84]) *Let G be a graph that does not have any isolated vertices. If G is bipartite and not a star, then $gr_*^t(G)$ is linear in t. If G is not bipartite, then $gr_*^t(G)$ is exponential in t.*

In 2021, Zhang et al. [94] considered Gallai-Ramsey numbers when the arguments are all copies of K_3 or $2K_3$. Their work built off of constructions established by Chung and Graham [18] in 1983, where it was shown that for all $t \geq 3$,

$$gr^t(K_3) = \begin{cases} 5^{t/2} + 1 & \text{if } t \text{ is even} \\ 2 \cdot 5^{(t-1)/2} + 1 & \text{if } t \text{ is odd.} \end{cases}$$

In order to describe their construction, begin by letting G_1 denote the 2-colored K_5 in which the subgraph spanned by each of the two colors is a C_5 (see the first image in Fig. 3.3).[2] For each $i \geq 2$, define G_i to be the $2i$-colored K_{5^i} formed by replacing each of the vertices in G_{i-1} (which uses colors $1, 2, \ldots, 2i - 2$) with a 2-colored K_5 in colors $2i - 1$ and $2i$ in which each of the colors spans a subgraph isomorphic to C_5. For example, G_2 is shown as the second image in Fig. 3.3. The resulting $2i$-colored K_{5^i} avoids a monochromatic K_3. Letting $t = 2i$, it follows that $gr^t(K_3) \geq 5^{t/2} + 1$ when t is even.

In the case where t is odd, let $t = 2i + 1 \geq 3$. Form G_i' by replacing each of the vertices in G_i (which uses colors $1, 2, \ldots, 2i$) with a K_2 in color $2i + 1$. For

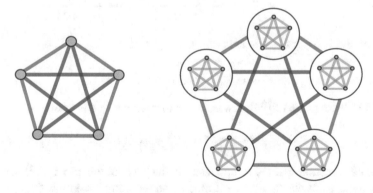

Fig. 3.3 A Gallai $2i$-coloring of K_{5^i} that avoids a monochromatic K_3, when $i = 1, 2$

---

[2] Figures 3.3, 3.4, 3.5, and 3.6 are reproduced from [3] with permission from *The Art of Discrete and Applied Mathematics*.

Fig. 3.4 A Gallai 3-coloring
of K_{10} that avoids a
monochromatic K_3

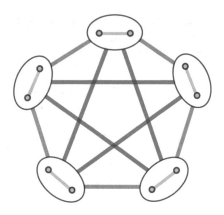

example, Fig. 3.4 shows the case $t = 3$. The resulting $2i + 1$-colored $K_{2 \cdot 5^i}$ avoids a monochromatic K_3, from which is follows that $gr^t(K_3) \geq 2 \cdot 5^{(t-1)/2} + 1$ when t is odd.

In the next theorem, the following simplifiied notation is used for the corresponding t-color Ramsey and star-critical Ramsey numbers. For each s such that $1 \leq s \leq t$, let

$$gr(s; t - s) := gr(\underbrace{2K_3, 2K_3, \ldots, 2K_3}_{s \text{ terms}}, \underbrace{K_3, K_3, \ldots, K_3}_{t-s \text{ terms}})$$

and

$$gr_*(s; t - s) := gr_*(\underbrace{2K_3, 2K_3, \ldots, 2K_3}_{s \text{ terms}}, \underbrace{K_3, K_3, \ldots, K_3}_{t-s \text{ terms}}).$$

Theorem 3.6 ([3]) *If $t \geq 4$ is even and $1 \leq s \leq t$, then*

$$gr_*(s; t - s) = gr(s; t - s) - 1.$$

Proof In Theorem 1.6 of [94], it was shown that when $t \geq 4$ is even,

$$gr(s; t - s) = 5^{t/2} + 2s + 1,$$

and the construction that led to this result is the basis of this proof. Following the construction given above, begin with the t-colored complete graph $G_{t/2}$ of order $5^{t/2}$ described above. This coloring was constructed from $G_{t/2-1}$ (which uses colors $1, 2, \ldots, t-2$) by replacing all of the vertices with copies of K_5 in colors $t-1$ and t in which each color spans a subgraph isomorphic to C_5. Select a block in $G_{t/2}$ (i.e., one of the K_5-subgraphs that replaced a vertex in $G_{t/2-1}$), and label it X. Next, select s vertices not contained in X and label them $x_1, x_2, \ldots, x_s$. This is possible since $s \leq t, t \geq 4$, and there are $5^{t/2} - 5 \geq t$ vertices to choose from.

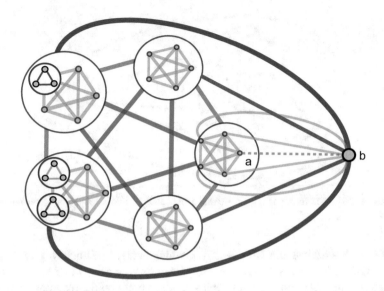

Fig. 3.5 A Gallai 4-coloring of $K_{32} - e$ that avoids a monochromatic copy of $2K_3$ in colors 1, 2, 3 and a monochromatic copy of K_3 in color 4

For each $1 \leq i \leq s$, replace x_i with a copy of K_3 in color i. In this construction, $2s$ vertices have been added to $G_{t/2}$, and while there exists K_3 subgraphs in each of the colors $1, 2, \ldots, s$, no $2K_3$-subgraph exists in any of these colors. Also, monochromatic K_3-subgraphs are still avoided in colors $s + 1, s + 2, \ldots, t$. The resulting t-coloring of $K_{5^{t/s}+2s}$ provides the lower bound for $gr(s; t - s)$.

Select a vertex $a \in X$ and introduce a new vertex b. For every $x \neq a, b$, color edge bx the same color as edge ax, leading to The edge joining a and b is the missing edge. For example, the construction corresponding to $s = 3$ and $t = 4$ is given in Fig. 3.5. The resulting $K_{5^{t/2}+2s+1} - e$ still lacks a monochromatic K_3 in colors $s + 1, s + 2, \ldots, t$ since any such K_3 created by the addition of b would have to include b, but no such subgraph existed that included a. For $1 \leq i \leq s$, every monochromatic K_3 in color i must include at least two vertices from the the K_3-subgraph that replaced vertex x_i. This prevents the existence of a $2K_3$ in color i. It follows that $gr_*(s; t - s) = 5^{t/2} + 2s$ and $(2K_3, 2K_3, \ldots, 2K_3, K_3, K_3, \ldots, K_3)$ is Gallai-Ramsey-full in this case.   □

Theorem 3.7 ([3]) *If $t \geq 3$ is odd and $k \in \mathbb{N}$ then the t-color star-critical Gallai-Ramsey number satisfies*

$$gr_*(K_3, K_3, \ldots, K_3, kK_3) = 2 \cdot 5^{(t-1)/2} + 3k - 3.$$

Proof This proof is based on Corollary 1.5 of [94], where it was shown that

$$gr(K_3, K_3, \ldots, K_3, kK_3) = 5^{(t-1)/2} + 3k - 2,$$

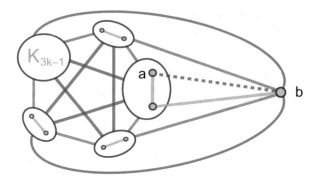

Fig. 3.6 A Gallai 3-coloring of $K_{3k+8} - e$ that avoids a K_3 in colors 1 and 2 and avoids a kK_3 in color 3

when $t \geq 3$ is odd. Begin with $G_{(t-1)/2}$, which is a $(t-1)$-coloring of $K_{5^{(t-1)/2}}$ using colors $1, 2, \ldots, t-1$. Replace a single vertex with a copy of K_{3k-1} in color t and replace all other vertices with copies of K_2 in color t. The resulting t-coloring of $K_{2 \cdot 5^{(t-1)/2}+3k-3}$ avoids a monochromatic copy of K_3 in colors $1, 2, \ldots, t-1$ and a monochromatic copy of kK_3 in color t.

Select a vertex in one of the K_2-subgraphs in color t and label it a. Introduce a vertex b and for all vertices $x \neq a, b$, color edge bx the same color as edge ax. The edge joining a and b is the missing edge. For example Fig. 3.6 shows the case $t = 3$. No monochromatic K_3 in colors $1, 2, \ldots t-1$ exists that includes vertex b since no such graph existed that included a. Since b only joins to vertices in the K_{3k-1} in color t via edges in colors other than t, no copy of kK_3 exists in color t. It follows that $gr_*(K_3, K_3, \ldots, K_3, kK_3) = 2 \cdot 5^{(t-1)/2} + 3k - 3$ and $(K_3, K_3, \ldots, K_3, kK_3)$ is Gallai-Ramsey-full in this case. □

Theorems 3.6 and 3.7 led the authors to state the following conjecture (see Open Problem 6).

Conjecture 3.2 ([3]) Let $m_i \in \mathbb{N}$ and $n_i \geq 3$ for all $1 \leq i \leq t$. Then

$$gr_*(m_1 K_{n_1}, m_2 K_{n_2}, \ldots, m_t K_{n_t}) = gr(m_1 K_{n_1}, m_2 K_{n_2}, \ldots, m_t K_{n_t}) - 1.$$

3.2 Deleting Subgraphs Other Than Stars

A recent variation of star-critical Ramsey numbers has been developed by Yan Li, Yusheng Li, and Ye Wang in a collection of papers ([62–65, 86, 87], and [88]). The idea they introduced involves deleting subgraphs other than stars from complete graphs to destroy the Ramsey property. Let $\mathbb{G} = \{H_k, H_{k+1}, \ldots\}$ be a class of graphs, where $k \geq 1$ and $\delta(H_n) \geq 1$ for $n \geq k$. Then for graphs G_1 and G_2 with $r(G_1, G_2) = p$, define the $\mathbb{G}$-*critical Ramsey number* $r_{\mathbb{G}}(G_1, G_2)$ to be the maximum number n such that every red/blue coloring of $K_p - E(H_n)$ (with $|V(H_n)| \leq p$) contains a red copy of G_1 or a blue copy of G_2.

For example, some classes of graphs worthy of consideration include

$$\mathbb{S} = \{K_{1,1}, K_{1,2}, \ldots\}, \quad \mathbb{M} = \{M_1, M_2, \ldots\}, \quad \mathbb{P} = \{P_2, P_3, \ldots\},$$

$$\text{and} \quad \mathbb{K} = \{K_2, K_3, \ldots\},$$

where $K_{1,n}$ is a star of size n, M_n is a matching of size n, P_n is a path of order n, and K_n is a complete graph of order n. In the case of $\mathbb{S}$, observe that

$$r_{\mathbb{S}}(G_1, G_2) = r(G_1, G_2) - 1 - r_*(G_1, G_2)$$

(equivalently, $r_{\mathbb{S}}(G_1, G_2) = de(G_1, G_2) - 1$). As such, results concerning the $\mathbb{S}$-critical Ramsey number have all been described in terms of star-critical Ramsey numbers in this text. These include results from [61–65, 87], and [88].

For $\mathbb{K}$-critical Ramsey numbers, Wang and Li [86] proved the following general upper bound.

Theorem 3.8 ([86]) *Let G_1 be a connected graph with order at least 2. Then*

$$r_{\mathbb{K}}(G_1, G_2) \leq \alpha(G_1) + r(G_1, G_2) - (|V(G_1)| - 1)(\chi(G_2) - 1) - 1.$$

In particular, if G_1 is G_2-good, then

$$r_{\mathbb{K}}(G_1, G_2) \leq \alpha(G_1) + s(G_2) - 1.$$

Proof First consider the case where $\chi(G_2) = 1$. In this case, G_2 is an empty graph (it does not contain any edges), from which it follows that $r(G_1, G_2) = |V(G_2)|$. Then

$$r_{\mathbb{K}}(G_1, G_2) = r(G_1, G_2) = |V(G_2)| \leq \alpha(G_1) + |V(G_2)| - 1,$$

corresponding with the first claim in the lemma.

Now assume that $\chi(G_2) \geq 2$ and let

$$t = \alpha(G_1) + r(G_1, G_2) - (|V(G_1)| - 1)(\chi(G_2) - 1) - 1.$$

The first claim in the lemma will follow from constructing a red/blue coloring of $K_{r(G_1,G_2)} - E(K_t)$ that contains a red copy of G_1 or a blue copy of G_2. Start with a blue $K_{\chi(G_2)-1}$ and replace one of its vertices with a red $K_{|V(G_1)|-\alpha(G_1)-1}$ and the other $\chi(G_2) - 2$ of its vertices with red $K_{|V(G_1)|-1}$-subgraphs. Introduce t vertices, forming a tK_1, and join them to the red $K_{|V(G_1)|-1-\alpha(G_1)-1}$-subgraph with red edges. Join them to the vertices in the red $K_{|V(G_1)|-1}$-subgraphs with blue edges. These t vertices form the deleted K_t-subgraph (see Fig. 3.7).

Fig. 3.7 A red/blue coloring of $K_{r(G_1, G_2)} - E(K_t)$ that avoids a red G_1 and a blue G_2

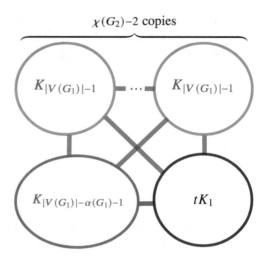

$\chi(G_2)-2$ copies

There are not enough vertices to contain the connected graph G_1 in any of the red $K_{|V(G_1)|-1}$-subgraphs. The other red component is isomorphic to $K_{|V(G_1)|-\alpha(G_1)-1} + tK_1$. Consider any selection of $|V(G_1)|$ vertices within this component. At least $\alpha(G_1) + 1$ of the vertices must come from the tK_1, and hence, the subgraph induced by the set of $|V(G_1)|$ vertices must have independence number exceeding that of G_1. Thus, no red G_1 is contained in this component either. Also, the subgraph spanned by the blue edges can be properly vertex colored by assigning colors according to which red component they are in. Hence, the chromatic number of any blue subgraph is at most $\chi(G_2) - 1$, preventing the existence of a blue G_2. It follows that

$$r_{\mathbb{K}}(G_1, G_2) \le \alpha(G_1) + r(G_1, G_2) - (|V(G_1)| - 1)(\chi(G_2) - 1) - 1.$$

Finally, if G_1 is G_2-good, then

$$r(G_1, G_2) = (|V(G_1)| - 1)(\chi(G_2) - 1) + s(G_2).$$

Plugging this into the above inequality results in

$$r_{\mathbb{K}}(G_1, G_2) \le \alpha(G_1) + s(G_2) - 1,$$

completing the proof of the theorem.   □

The inequality in Theorem 3.8 is similar to the 2-color case of Theorem 1.3, if it were stated in terms of $r_{\mathbb{S}}(G_1, G_2)$ rather than in terms of $r_*(G_1, G_2)$. This result is useful in determining the value of $r_{\mathbb{K}}(K_{1,m}, K_n)$.

Theorem 3.9 ([86]) *For all integers $m, n \ge 2$, $r_{\mathbb{K}}(K_{1,m}, K_n) = m$.*

Proof In [21], Chvátal proved that $r(K_{1,m}, K_n) = m(n-1) + 1$. Note that $K_{1,m}$ is n-good. Theorem 3.8 then implies that $r_{\mathbb{K}}(G_1, G_2) \leq \alpha(K_{1,m}) = m$.

To prove the reverse inequality, it must be shown that every red/blue coloring of $K_{m(n-1)+1} - E(K_m)$ contains a red $K_{1,m}$ or a blue K_n. If $n = 2$, then $K_{m+1} - E(K_m) = K_{1,m}$. There exists a blue K_2 if any edge is blue. If all of the edges are red, then the graph forms is a red $K_{1,m}$. It follows that $r_{\mathbb{K}}(K_{1,m}, K_2) \geq m$.

For $n \geq 3$, consider a red/blue coloring of $K_{m(n-1)+1} - E(K_m)$ that avoids a blue K_n. Let $y_1, y_2, \ldots, y_{m(n-2)+1}$ be vertices that are not in the K_m whose edges are deleted. Since $r(K_{1,m}, K_{n-1}) = m(n-2) + 1$, the subgraph induced by $\{y_1, y_2, \ldots, y_{m(n-2)+1}\}$ contains a red $K_{1,m}$ or a blue K_{n-1}. Assume the latter. Since a blue K_n is avoided, every vertex not in the blue K_{n-1} joins to at least one vertex in the K_{n-1} with a red edge. As there are $(m-1)(n-1) + 1$ such vertices, by the pigeonhole principle, at least one of the vertices in the blue K_{n-1} must join to at least m other vertices via red edges, forming a red $K_{1,m}$. It follows that $r_{\mathbb{K}}(K_{1,m}, K_n) \geq m$ when $n \geq 3$, completing the proof.   □

This section is concluded by listing two additional theorems by Wang and Li [86] concerning $\mathbb{K}$-critical Ramsey numbers. The deletion of other classes of graphs than $\mathbb{S}$ and $\mathbb{K}$ is mostly unexplored at present (see Open Problem 9).

Theorem 3.10 ([86]) *For any integer $n \geq 4$, $r_{\mathbb{K}}(C_4, P_n) = \lceil \frac{n}{2} \rceil$.*

Theorem 3.11 ([86]) *For any integer $n \geq 2$, $r_{\mathbb{K}}(K_3, F_n) = n$.*

Chapter 4
Directions for Future Research and Summary of Known Values

4.1   Directions for Future Study

Some time dedicated to a careful inspection of Radziszowski's dynamic survey
[76] can uncover numerous known Ramsey numbers whose star-crtical counterparts
have not yet been investigated. Besides these open cases, this section includes a
list of some open problems that have naturally presented themselves through the
material in this book. It is hoped that the following open problems will provide
good starting points for continued research into star-critical Ramsey numbers, and
their generalizations.

1. Let $m_1, m_2, \ldots, m_t$ be integers greater than 1, exactly k of which are even.
   Then Burr and Roberts [14] proved that

$$r(K_{1,m_1}, K_{1,m_2}, \ldots, K_{1,m_t}) = \begin{cases} N - t + 1 \text{ if } k \geq 2 \text{ is even} \\ N - t + 2 \text{ otherwise,} \end{cases}$$

   where

$$N = \sum_{i=1}^{t} m_i.$$

   In Theorems 2.4 and 2.5, the corresponding star-critical Ramsey number was
   evaluated in the cases where k is 0, 2, or is odd (see [5]). At present, its value
   has not been determined when $k > 2$ is even, although it is conjectured (see
   Conjecture 2.1) to be

$$r_*(K_{1,m_1}, K_{1,m_2}, \ldots, K_{1,m_t}) = N - t,$$

   in this case.

© The Author(s), under exclusive license to Springer Nature Switzerland AG 2023   81
M. R. Budden, *Star-Critical Ramsey Numbers for Graphs*, SpringerBriefs in
Mathematics, https://doi.org/10.1007/978-3-031-29981-0_4

2. In Theorem 2.8, the upper bound $r_*(T_m^*, T_n^*) \leq m + n - 5$ was proved for all $m, n \geq 5$ that satisfy either $(m - 1)|(n - 3)$ or $(n - 1)|(m - 3)$. Can any improvements be made to this upper bound? Can a corresponding lower bound also be determined? Besides the case that restricts to the given divisibility properties, one can also consider $r_*(T_m^*, T_n^*)$ in other cases. See Theorem 3.2 in Guo and Volkmann's paper [38] for the Ramsey numbers for other cases.

3. From Theorem 2.7, $r_*(P_m, K_{1,n})$ has been evaluated whenever $m \geq 2n + 1$. What about the cases where $m < 2n + 1$? Note that Theorem 2.6 gives an upper bound for $r_*(P_m, K_{1,n})$ when $(m - 1)|(n - 1)$. Is this bound exact?

4. Theorem 2.28 gave the evaluation of $r_*(K_3, W_n)$ when $n \geq 7$. The case $n = 4$ corresponds with $r_*(K_3, K_4) = 8$, which follows from Theorem 1.1. The values of $r_*(K_3, W_5)$ and $r_*(K_3, W_6)$ have not yet been determined.

5. In [84], Su and Liu conjectured that if G is a graph that does not contain any isolated vertices, then G is Ramsey-full if and only if G is Gallai-Ramsey-full (see Conjecture 3.1). It was noted in Sect. 3.1 that this conjecture is true for complete graphs. Besides those considered by Su and Liu, for what other graphs can this conjecture be verified?

6. Conjecture 3.2 in Sect. 3.1 states that $(m_1 K_{n_1}, m_2 K_{n_2}, \ldots, m_t K_{n_t})$ is Gallai-Ramsey-full for all $m_i \in \mathbb{N}$ and $n_i \geq 3$, where $1 \leq i \leq t$. Theorems 3.6 and 3.7 offer support for this conjecture and it may be possible to use additional results from Zhang et al.'s paper [94] to prove additional cases where $n_i = 3$ for all i.

7. Weakened Ramsey numbers were first introduced by Chung et al. in 1977 [19]. For $1 \leq s < t$, define the *weakened Ramsey number* $r_s^t(G)$ to be the least natural number p such that every t-coloring of K_p contains a subgraph isomorphic to G that is spanned by edges using at most s colors. From this definition, it follows that $r_1^t(G) = r^t(G)$. Currently, the star-critical Ramsey number for weakened Ramsey numbers has not been studied when $s \geq 2$.

8. One can further generalize the concept of a weakened Ramsey number by restricting to Gallai colorings (see [2] and [7]). The *weakened Gallai-Ramsey number* $gr_s^t(G)$ is the least natural number p such that every Gallai t-coloring of K_p contains a subgraph isomorphic to G that is spanned by edges using at most s colors, where $1 \leq s < t$. Currently, the star-critical Ramsey number for weakened Gallai-Ramsey numbers has not been studied when $s \geq 2$.

9. The $\mathbb{G}$-critical Ramey numbers defined by Li et al. [62–65, 86, 87], and [88] can be defined and studied for other classes of graphs than those given in Sect. 3.2. For example, one can consider the classes

$$\mathbb{B} = \{K_{1,1}, K_{2,2}, \ldots\}, \quad \mathbb{C} = \{C_3, C_4, \ldots\} \quad \text{and} \quad \mathbb{D} = \{K_3 - e, K_4 - e, \ldots\}.$$

10. For $r \geq 2$, an r-*uniform hypergraph* H consists of a vertex set $V(H)$ and a set of r-uniform hyperedges $E(H)$, whose elements are r-element subsets of $V(H)$. The *complete r-uniform hypergraph of order n*, denoted $K_n^{(r)}$, consists of n vertices in which every r-element subset is a hyperedge. Analogous to the Ramsey number for graphs, one can define the Ramsey number $r(H_1, H_2, \ldots, H_t; r)$,

where $H_1, H_2, \ldots, H_t$ are r-uniform hypergraphs. The most natural way to extend star-critical Ramsey theory to hypergraphs follows the generalization of Li, Li, and Wang described in Sect. 3.2, by deleting subhypergraphs of a certain class. For example, for $n \geq r$, let $S_n^{(r)}$ be the r-*uniform star*, defined to have center vertex x, $n-1$ other vertices $y_1, y_2, \ldots, y_{n-1}$, and whose hyperedges are those that contain x and $r-1$ vertices from $\{y_1, y_2, \ldots, y_{n-1}\}$. For the classes of hypergraphs

$$\mathbb{S}^{(r)} = \{S_r^{(r)}, S_{r+1}^{(r)}, \ldots\} \quad \text{and} \quad \mathbb{K}^{(r)} = \{K_r^{(r)}, K_{r+1}^{(r)}, \ldots\},$$

one can define $\mathbb{S}^{(r)}$ and $\mathbb{K}^{(r)}$-critical Ramsey numbers analogously to $\mathbb{G}$-critical Ramsey numbers.

4.2 Summary of Bounds and Tables of Known Values

In this section, the current known (nonasymptotic) evaluations of star-critical Ramsey numbers are summarized for easy reference in Tables 4.1, 4.2, 4.3, 4.4, 4.5, and 4.6. In the case of complete graphs,

$$r_*(K_{n_1}, K_{n_2}, \ldots, K_{n_t}) = r(K_{n_1}, K_{n_2}, \ldots, K_{n_t}) - 1,$$

where $n_i \geq 2$ for all $1 \leq i \leq t$ (see Theorem 1.1). The Gallai version of this result also holds. If $n_i \geq 2$ and $1 \leq i \leq t$, then

$$gr_*(K_{n_1}, K_{n_2}, \ldots, K_{n_t}) = gr(K_{n_1}, K_{n_2}, \ldots, K_{n_t}) - 1.$$

Recall that P_n is a path of order n, T_n is a tree of order n, F is a forest, $K_{1,n}$ is a star of order $n+1$, K_n is a complete graph of order n, mK_n is the disjoint union of m copies of K_n, C_n is a cycle of order n, W_n is a wheel of order n, $F_n = K_1 + nK_2$, $B_n = K_2 + nK_1$, $F_{t,n} = K_1 + nK_t$, and $B_p(n) = K_p + nK_1$.

Table 4.1 Known star-critical Ramsey numbers involving trees

Star-critical Ramsey number	Location in this book
$r_*(P_m, P_n) = \lceil \frac{m}{2} \rceil$, for all $n \geq m \geq 2$ [47, 48]	Theorem 2.2
$r_*^t(P_3) = 1$, for all odd $t \geq 1$. [5]	Theorem 2.3
$r_*(K_{1,m_1}, K_{1,m_2}, \ldots, K_{1,m_t})) = \begin{cases} N - t & \text{if } k = 2 \\ 1 & \text{if } k = 0 \text{ or } k \text{ is odd,} \end{cases}$ where N is the sum of the m_i and k is the number of m_i that are even. [5]	Theorems 2.4 and 2.5
$r_*(P_m, K_{1,n}) = n$, for all $m, n \in \mathbb{N}$ such that $m \geq 2n + 1$. [88]	Theorem 2.7
$r_*(T_m, K_{n_1}, K_{n_2}, \ldots, K_{n_t}) = (m - 1)(r(K_{n_1}, K_{n_2}, \ldots, K_{n_t}) - 2) + 1$, for all $m, n_i \geq 2$, where $1 \leq i \leq t$. [5, 47, 50]	Theorem 2.10
$r_*(F, K_{n_1}, K_{n_2}, \ldots, K_{n_t}) = (j_0 - 1)(p - 3) + \sum_{i=j_0}^{m(F)} i k_i(F)$, where $p = r(K_{n_1}, K_{n_2}, \ldots, K_{n_t})$ and $j_0, k_i(F)$, and $m(F)$ are defined in Sect. 2.2. [55]	Theorem 2.11
$r_*(P_3, K_n - e) = 2n - 5$, for all odd $n \geq 4$. [47]	Theorem 2.12
$r_*(P_4, K_4 - e) = 4$. [5]	Theorem 2.13
$r_*(P_3, P_3, K_{n_1}, K_{n_2}, \ldots, K_{n_t}) = 2r(K_{n_1}, K_{n_2}, \ldots, K_{n_t}) - 2$, for all $n_i \geq 2$, where $1 \leq i \leq t$. [5]	Theorem 2.14
$r_*(C_4, P_n) = 3$, for all $n \geq 3$. [47, 50]	Theorem 2.22
$r_*(C_m, K_{1.n}) = n + 1$, for all $m, n \in \mathbb{N}$ such that $m \geq 2n + 2$. [88]	Theorem 2.23
$r_*(P_m, F_n) = m + n - 1$, for all $n \geq 2$ and $m \geq 2n + 1$. [88]	Theorem 2.40
$r_*(P_m, B_n) = m$, for all $n \geq 4$ and $m \geq 2n$. [88]	Theorem 2.47
$r_*(K_{1,m-1}, B_n) = m$, for all $m, n \in \mathbb{N}$ such that $m \geq 3n - 1$. [87]	Theorem 2.48

Table 4.2 Known star-critical Ramsey numbers involving cycles or wheels

Star-critical Ramsey number	Location in this book
$r_*(C_3, C_n) = n + 1$, for all $n \geq 3$. [47]	Theorem 2.16
$r_*(C_4, C_n) = 5$, for all $n \geq 4$. [90]	Theorems 2.17 and 2.19
$r_*(C_m, C_n) = n + 1$, for m odd, $n \geq m \geq 3$, and $(m, n) \neq (3, 3)$. [95]	Theorem 2.20
$r_*(C_4, P_n) = 3$, for all $n \geq 3$. [47, 50]	Theorem 2.22
$r_*(C_m, K_{1,n}) = n + 1$, for all $m, n \in \mathbb{N}$ such that $m \geq 2n + 2$. [88]	Theorem 2.23
$r_*(C_4, K_4) = 9$. [54]	Theorem 2.24
$r_*(C_m, K_4) = 2m$, for all $m \geq 5$. [47, 54]	Theorem 2.26
$r_*(C_3, W_n) = n + 2$, for all $n \geq 7$. [47]	Theorem 2.28
$r_*(C_m, W_n) = 2m$, for n even, $m + 1 \geq n \geq 6$, and $m \geq 60$. [67]	Theorem 2.30
$r_*(C_m, W_n) = n + 2$, for odd $m \geq 5$ and $n \geq \frac{3(m-1)}{2} + 3$. [68]	Theorem 2.31
$r_*(C_3, F_n) = 2n + 2$, for all $n \geq 2$. [66]	Theorem 2.42
$r_*(B_2, C_3) = 5$. [5]	Theorem 2.45
$r_*(C_m, B_n) = \begin{cases} 2m & \text{if } \frac{9n}{10} \leq m < n \\ 2n + 1 & \text{if } n + 1 \leq m \leq \frac{10n}{9}, \end{cases}$ for n large. [64]	Theorem 2.49
$r_*(K_{1,3} + e, C_3) = 4$. [5]	Theorem 2.51
$r_*(C_3, F_{3,n}) = 3n + 3$, for all $n \geq 4$. [42]	Theorem 2.53
$r_*(C_{2k+1}, B_p(n)) = r(C_{2k+1}, B_p(n)) - n$, for $k, p \in \mathbb{N}$ and n is large. [62]	Theorem 2.57
$r_*(C_{2k+1}, K_1 + nG) = r(C_{2k+1}, K_1 + nG) - |V(G)|n + \delta(G)$, where $k \in \mathbb{N}$ and n is large. [62]	Theorem 2.58

Table 4.3 Known star-critical Ramsey numbers involving disjoint unions of complete graphs

Star-critical Ramsey number	Location in this book
$r_*(mK_2, nK_2) = m$, for all $n \geq m \geq 1$. [47, 50]	Theorem 2.33
$r_*(m_1 K_2, m_2 K_2, \ldots, m_t K_2) = 1 + \sum\limits_{i=2}^{t} (m_i - 1)$, for all $m_1 \geq m_2 \geq \cdots \geq m_t \geq 1$. [47, 50, 92]	Theorems 2.33 and 2.34
$r_*(K_m, nK_2) = m + 2n - 3$, for all $m \geq 3$ and $n \geq 1$. [66]	Theorem 2.35
$r_*(mK_3, nK_3) = 2m + 3n - 1$, for all $n \geq 2$ and $n \geq m \geq 1$. [47, 50]	Theorem 2.37
$r_*(mK_3, nK_4) = \begin{cases} 2m + 4n & \text{if } n \geq m \\ 3m + 3n & \text{otherwise}, \end{cases}$ for all $m \geq 1$ and $n \geq 2$. [66]	Theorem 2.38
$r_*(mK_k, nK_\ell) = (k - 1)m + (\ell - 1)n + \max\{m, n\} + p - 3$, where $p = r(K_{k-1}, K_{\ell-1})$, $k, \ell \geq 3$, and m and n are large. [95]	Theorem 2.39

Table 4.4   Known star-critical Ramsey numbers involving fans or books

Star-critical Ramsey number	Location in this book
$r_*(P_m, F_n) = m + n - 1$, for all $n \geq 2$ and $m \geq 2n + 1$. [88]	Theorem 2.40
$r_*(K_3, F_n) = 2n + 2$, for all $n \geq 2$. [66]	Theorem 2.42
$r_*(K_4, F_n) = 4n + 2$, for all $n \geq 4$. [40]	Theorem 2.44
$r_*(B_2, K_3) = 5$. [5]	Theorem 2.45
$r_*(B_2, K_{1,3} + e) = 5$. [5]	Theorem 2.46
$r_*(P_m, B_n) = m$, for all $n \geq 4$ and $m \geq 2n$. [88]	Theorem 2.47
$r_*(K_{1,m-1}, B_n) = m$, for all $m, n \in \mathbb{N}$ such that $m \geq 3n - 1$. [87]	Theorem 2.48
$r_*(C_m, B_n) = \begin{cases} 2m & \text{if } \frac{9n}{10} \leq m < n \\ 2n + 1 & \text{if } n + 1 \leq m \leq \frac{10n}{9}, \end{cases}$ for n large. [64]	Theorem 2.49
$r_*(K_3, F_{3,n}) = 3n + 3$, for all $n \geq 4$. [42]	Theorem 2.53
$r_*(C_{2k+1}, B_p(n)) = r(C_{2k+1}, B_p(n)) - n$, for $k, p \in \mathbb{N}$ and n is large. [62]	Theorem 2.57

Table 4.5   Known t-color star-critical Gallai-Ramsey numbers

Star-critical Gallai-Ramsey number	Location in this book
$gr_*^t(C_4) = t + 3$, for all $t \geq 2$. [84]	Theorem 3.2
$gr_*^t(P_4) = t$, for all $t \in \mathbb{N}$. [84]	Theorem 3.3
$gr_*^t(K_{1,m}) = \begin{cases} 2m - 2 & \text{if } m \geq 12 \text{ is even} \\ m & \text{if } m \geq 3 \text{ is odd,} \end{cases}$ for all $t \geq 3$. [84]	Theorem 3.4
$gr_*(s; t - s) = gr(s; t - s) - 1$, for all even $t \geq 4$ and $t \geq s \geq 1$. [3]	Theorem 3.6
$gr_*(K_3, K_3, \ldots, K_3, kK_3) = 2 \cdot 5^{(t-1)/2} + 3k - 3$, for all odd $t \geq 3$ and $k \in \mathbb{N}$. [3]	Theorem 3.7

Table 4.6   Known $\mathbb{K}$-critical Ramsey numbers

Star-critical Gallai-Ramsey number	Location in this book
$r_{\mathbb{K}}(K_{1,m}, K_n) = m$, for all $m, n \geq 2$. [86]	Theorem 3.9
$r_{\mathbb{K}}(C_4, P_n) = \lceil \frac{n}{2} \rceil$, for all $n \geq 4$. [86]	Theroem 3.10
$r_{\mathbb{K}}(K_3, F_n) = n$, for all $n \geq 2$. [86]	Theorem 3.11

References

1. J. Bondy, P. Erdős, Ramsey number for cycles in graphs. J. Combin. Theory Ser. B **14**, 46–54 (1973)
2. G. Beam, M. Budden, Weakened Gallai-Ramsey numbers. Surv. Math. Appl. **13**, 131–145 (2018)
3. M. Budden, Z. Daouda, Star-critical Gallai-Ramsey numbers involving the disjoint union of triangles. Art Discrete Appl. Math. **6**, #P1.09 (2023)
4. M. Budden, E. DeJonge, Destroying the Ramsey property by the removal of edges. Australas. J. Combin. **74**, 476–485 (2019)
5. M. Budden, E. DeJonge, Multicolor star-critical Ramsey numbers and Ramsey-good graphs. Electron. J. Graph Theory Appl. **10**, 51–66 (2022)
6. M. Budden, J. Hiller, A. Penland, Constructive methods in Gallai-Ramsey theory for hypergraphs. Integers **20A**. #A4 (2020)
7. M. Budden, T. Wimbish, Subgraphs of Gallai-colored complete graphs spanned by edges using at most two colors. Australas. J. Combin. **84**, 375–387 (2022)
8. S. Burr, Generalized Ramsey theory for graphs – a survey. In *Graphs and Combinatorics* (Proc. Capital Conf., George Washington Univ., Washington, D.C., 1973), Lecture Notes in Math. **406**, 52–75 (1974)
9. S. Burr, Ramsey numbers involving graphs with long suspended paths. J. London Math. Soc. **24**(2), 405–413 (1981)
10. S. Burr, On Ramsey numbers for large disjoint unions of graphs. Discrete Math. **70**, 277–293 (1988)
11. S. Burr, P. Erdős, Generalizations of a Ramsey-theoretic result of Chvátal. J. Graph Theory **7**, 39–51 (1983)
12. S. Burr, P. Erdős, R. Faudree, C. Rousseau, R. Schelp, Some complete bipartite graph-tree Ramsey numbers. Ann. Discrete Math. **41**, 79–89 (1989)
13. S. Burr, P. Erdős, J. Spencer, Ramsey theorems for multiple copies of graphs. Trans. Amer. Math. Soc. **209**, 87–99 (1975)
14. S. Burr, J. Roberts, On Ramsey numbers for stars. Utilitas Math. **4**, 217–220 (1973)
15. G. Chartrand, L. Lesniak, P. Zhang, *Graphs and Digraphs, Fifth Edition* (Taylor & Francis Group, Boca Raton, 2011)
16. G. Chartrand, S. Schuster, On a variation of the Ramsey number. Trans. Amer. Math. Soc. **173**, 353–362 (1972)
17. Y. Chen, T. Cheng, C. Ng, Y. Zhang, A theorem on cycle-wheel Ramsey number. Discrete Math. **312**, 1059–1061 (2012)

© The Author(s), under exclusive license to Springer Nature Switzerland AG 2023
M. R. Budden, *Star-Critical Ramsey Numbers for Graphs*, SpringerBriefs in Mathematics, https://doi.org/10.1007/978-3-031-29981-0

18. F. Chung, R. Graham, Edge-colored complete graphs with precisely colored subgraphs. Combinatorica **3**, 315–324 (1983)
19. K. Chung, M. Chung, C. Liu, A generalization of Ramsey theory for graphs – with stars and complete graphs as forbidden subgraphs. Congr. Numer. **19**, 155–161 (1977)
20. V. Chvátal, On Hamilton's ideals. J. Combin. Theory Ser. B **12**, 163–168 (1972)
21. V. Chvátal, Tree-complete graph Ramsey numbers. J. Graph Theory **1**, 93 (1977)
22. V. Chvátal, F. Harary, Generalized Ramsey theory for graphs III. Small off-diagonal numbers. Pacific J. Math. **41**, 335–345 (1972)
23. E. Cockayne, P. Lorimer, On Ramsey graph numbers for stars and stripes. Canad. Math. Bull. **18**, 31–34 (1975)
24. E. Cockayne, P. Lorimer, The Ramsey number for stripes. J. Austral. Math. Soc. (Series A) **19**, 252–256 (1975)
25. G. Dirac, Some theorems on abstract graphs. Proc. Lond. Math. Soc. **2**, 68–81 (1952)
26. P. Erdős, R. Faudree, Size Ramsey functions. In *Sets, Graphs and Numbers* (Budapest, 1991), Colloq. Math. Soc. Janos Bolyai **60**, 219–238 (1992)
27. P. Erdős, R. Faudree, C. Rousseau, R. Schelp, The size Ramsey number. Period. Math. Hungar. **9**, 145–161 (1978)
28. P. Erdős, R. Faudree, C. Rousseau, R. Schelp, On cycle-complete graph Ramsey numbers. J. Graph Theory **2**, 53–64 (1978)
29. P. Erdős, R. Faudree, C. Rousseau, R. Schelp, The book-tree Ramsey numbers. Scientia, Ser. A, Math. Sci. **1**, 111–117 (1988)
30. R. Faudree, R. Gould, M. Jacobson, C. Magnant, Ramsey numbers in rainbow triangle free colorings. Australas. J. Combin. **46**, 269–284 (2010)
31. R. Faudree, S. Lawrence, T. Parsons, R. Schelp, Path-cycle Ramsey numbers. Discrete Math. **10**, 269–277 (1974)
32. R. Faudree, R. Schelp, All Ramsey numbers for cycles in graphs. Discrete Math. **8**, 313–329 (1974)
33. T. Gallai, Transitiv orientierbare graphen. Acta Math. Acad. Sci Hungar. **18**, 25–66 (1967)
34. L. Gerencsér, A. Gyárfás, On Ramsey-type problems. Ann. Univ. Sci. Budapest. Eötvös Sect. Math. **10**, 167–170 (1967)
35. R. Graham, S. Butler, *Rudiments of Ramsey Theory, Second Edition* (Regional Conference Series in Mathematics, Number 123, American Mathematical Society, Providence, Providence, 2015)
36. R. Graham, B. Rothshild, J. Spencer, *Ramsey theory* (John Wiley & Sons, New York, 1980)
37. R. Greenwood, A. Gleason, Combinatorial relations and chromatic graphs. Canad. J. Math. **7**, 1–7 (1955)
38. Y. Guo, L. Volkmann, Tree-Ramsey numbers. Australas. J. Combin. **11**, 169–175 (1995)
39. A. Gyárfás, G. Simonyi, Edge colorings of complete graphs without tricolored triangles. J. Graph Theory **46**(3), 211–216 (2004)
40. S. Haghi, H. Maimani, A. Seify, Star-critical Ramsey numbers of F_n versus K_4. Discrete Appl. Math. **217**, 203–209 (2017)
41. A. Hamm, P. Hazelton, S. Thompson, On Ramsey and star-critical Ramsey numbers for generalized fans versus nK_m. Discrete Appl. Math. **305**, 64–70 (2021)
42. Y. Hao, Q. Lin, Ramsey number of K_3 versus $F_{3,n}$. Discrete Appl. Math. **251**, 345–348 (2018)
43. Y. Hao, Q. Lin, Star-critical Ramsey numbers for large generalized fans and books. Discrete Math. **341**, 3385–3393 (2018)
44. F. Harary, *Graph Theory* (Addison-Wesley Publishing Co., Reading, 1969)
45. F. Harary, Recent results on generalized Ramsey theory for graphs. In *Graph Theory and Applications* (Proc. Conf., Western Michigan Univ., 1972) Lecture Notes in Math. **303**, 125–138 (1972)
46. F. Harary, G. Prins, Generalized Ramsey theory for graphs, IV: the Ramsey multiplicity of a graph. Networks **4**, 163–173 (1974)
47. J. Hook, *The Classification of Critical Graphs and Star-Critical Ramsey Numbers* (Ph.D. Thesis, Lehigh Univ., 2010)

48. J. Hook, Critical graphs for $R(P_n, P_m)$ and the star-critical Ramsey number for paths. Discuss. Math. Graph Theory **35**, 689–701 (2015)
49. J. Hook, *Recent developments of star-critical Ramsey numbers* Accepted - Springer Proceedings in Mathematics & Statistics: Combinatorics, Graph Theory, and Computing. SEICGTC 2021)
50. J. Hook, G. Isaak, Star-critical Ramsey numbers. Discrete Appl. Math. **159**, 328–334 (2011)
51. R. Irving, Generalised Ramsey numbers for small graphs. Discrete Math. **9**, 251–264 (1974)
52. M. Jacobson, On the Ramsey multiplicity for stars. Discrete Math. **42**, 63–66 (1982)
53. M. Jacobson, On the Ramsey number for stars and a complete graph. Ars Combin. **17**, 167–172 (1984)
54. C. Jayawardene, D. Narváez, S. Radziszowski, Star-critical Ramsey numbers for cycles versus K_4. Discuss. Math. Graph Theory **41**, 381–390 (2021)
55. A. Kamranian, G. Raeisi, On the star-critical Ramsey number of a forest versus complete graphs. Iran. J. Sci. Technol. Trans. Sci. **46**, 499–505 (2022)
56. M. Katz, J. Reimann, *An Introduction to Ramsey Theory – Fast Functions, Infinity, and Metamathematics* (Student Mathematical Library, Volume 87, American Mathematical Society, Providence, 2018)
57. B. Landman, A. Robertson, *Ramsey Theory on the Integers, Second Edition* (Student Mathematical Library, Volume 73, American Mathematical Society, Providence, 2014)
58. S. Lawrence, Cycle-star Ramsey numbers. Notices Amer. Math. Soc. **20**, Abstract A-420 (1973)
59. Y. Li, C. Rousseau, Fan-complete Ramsey numbers. J. Graph Theory **23**(4), 413–420 (1996)
60. Y. Li, Q. Lin, *Elementary Methods of Graph Ramsey Theory* (Applied Mathematical Sciences, Volume 211, Springer, Cham, 2022)
61. Y. Li, Y. Li, Star-critical Ramsey numbers involving large books. Discrete Appl. Math. **327**, 68–76 (2023)
62. Y. Li, Y. Li, Y. Wang, Minimal Ramsey graphs on deleting stars for generalized fans and books. Appl. Math. Comput. **372**, #125006 (2020)
63. Y. Li, Y. Li, Y. Wang, Star-critical Ramsey numbers of generalized fans. Graphs Combin. **37**, 2113–2120 (2021)
64. Y. Li, Y. Li, Y. Wang, Star-critical Ramsey number of large cycle and book of different orders. Theoret. Comput. Sci. **866**, 37–42 (2021)
65. Y. Li, Y. Li, Y. Wang, Star-critical Ramsey numbers involving large generalized fans. Graphs Combin. **38**. #130 (2022)
66. Z. Li, Y. Li, Some star-critical Ramsey numbers. Discrete Appl. Math. **181**, 301–305 (2015)
67. Y. Liu, Y. Chen, Star-critical Ramsey numbers of cycles versus wheels. Graphs Combin. **37**, 2167–2172 (2021)
68. Y. Liu, Y. Chen, Star-critical Ramsey numbers of wheels versus odd cycles. Acta Math. Appl. Sin. Engl. Ser. **38**, 916–924 (2022)
69. P. Lorimer, The Ramsey numbers for stripes and one complete graph. J. Graph Theory **8**, 177–184 (1984)
70. P. Lorimer, P. Mullins, Ramsey numbers for quadrangles and triangles. J. Combin. Theory Ser. B **23**, 262–265 (1977)
71. C. Magnant, P. Salehi Nowbandegani, *Topics in Gallai-Ramsey Theory* (SpringerBriefs in Mathematics, Springer, Cham, 2020)
72. J. Nešetřil, V. Rödl (Editors), *Mathematics of Ramsey Theory* (Algorithms and Combinatorics, Volume 5, Springer-Verlag, Berlin, 1990)
73. G. Omidi, G. Raeisi, A note on the Ramsey number of stars-complete graphs. Eur. J. Combin. **32**, 598–599 (2011)
74. Ø. Ore, Hamiltonian connected graphs. J. Math. Pure Appl. **42**, 21–27 (1963)
75. T. Parsons, Path-star Ramsey numbers. J. Combin. Theory Ser. B **17**, 51–58 (1974)
76. S. Radziszowski, Small Ramsey numbers – revision 16. Electron. J. Combin. **DS1.16**, 116 pages (2021)

77. S. Radziszowski, X. Jin, Paths, cycles and wheels in graphs without antitriangles. Australas. J. Combin. **9**, 221–232 (1994)
78. J.L. Ramírez-Alfonsín, B. Reed, *Perfect Graphs* (John Wiley & Sons, New York, 2001)
79. A. Robertson, *Fundamentals of Ramsey Theory* (Discrete Mathematics and Its Applications, Taylor & Francis Group, Boca Raton, 2021)
80. C. Rousseau, J. Sheehan, A class of Ramsey problems involving trees. J. Lond. Math. Soc. **18**, 392–396 (1978)
81. A. Salman, H. Broersma, Path-fan Ramsey numbers. Discrete Appl. Math. **154**, 1429–1436 (2006)
82. A. Soifer (Editor), *Ramsey Theory – Yesterday, Today, and Tomorrow* (Progress in Mathematics, Volume 285, Springer Science + Business Media, LLC, New York, 2011)
83. S. Stahl, On the Ramsey number $R(F, K_m)$ where F is a forest. Can. J. Math. **27**, 585–589 (1975)
84. X. Su, Y. Liu, Star-critical Gallai-Ramsey numbers of graphs. Graphs Combin. **38**, #158 (2022)
85. Surahmat, E. Baskoro, H. Broersma, The Ramsey numbers of fans versus K_4. Bull. Inst. Combin. Appl. **43**, 96–102 (2005)
86. Y. Wang, Y. Li, Deleting edges from Ramsey graphs. Discrete Math. **343**, #111743 (2020)
87. Y. Wang, Y. Li, Y. Li, Maximum star deleted from Ramsey graphs of book and tree. Theoret. Comput. Sci. **836**, 37–41 (2020)
88. Y. Wang, Y. Li, Y. Li, Star-critical Ramsey numbers involving graphs with long suspended paths. Discrete Math. **344**. #112233 (2021)
89. D. West, *Introduction to Graph Theory, Second Edition* (Prentice Hall, Upper Saddle River, 2000)
90. Y. Wu, Y. Sun, S. Radziszowski, Wheel and star-critical Ramsey numbers for quadrilateral. Discrete Appl. Math. **186**, 260–271 (2015)
91. X. Xu, M. Liang, H. Luo, *Ramsey Theory - Unsolved Problems and Results* (De Gruyter, Berlin, 2018)
92. C. Xu, H. Yang, S. Zhang, On characterizing the critical graphs for matching Ramsey numbers. Discrete Appl. Math. **287**, 15–20 (2020)
93. J. Yang, Y. Huang, K. Zhang, The value of the Ramsey number $R(C_n, K_4)$ is $3(n-1)+1$ ($n \geq 4$). Australas. J. Combin. **20**, 205–206 (1999)
94. F. Zhang, X. Zhu, Y. Chen, Gallai-Ramsey numbers for multiple triangles. Discrete Appl. Math. **298**, 103–109 (2021)
95. Y. Zhang, H. Broersma, Y. Chen, On star-critical and upper size Ramsey numbers. Discrete Appl. Math. **202**, 174–180 (2016)

Index

Printed in the United States
by Baker & Taylor Publisher Services